Introduction to Immunocytochemistry

THIRD EDITION

Introduction to Immunocytochemistry

THIRD EDITION

J.M. Polak
Director, Tissue Engineering and
Regenerative Medicine Centre,
Imperial College Faculty of Medicine,
Chelsea & Westminster Hospital,
London, UK

S. Van Noorden
Department of Histopathology,
Imperial College Faculty of Medicine,
Hammersmith Hospital,
London, UK

© BIOS Scientific Publishers Limited, 2003

First edition © Royal Microscopical Society, 1984, 1987
Second edition © BIOS Scientific Publishers, 1997
Third edition published 2003

A CIP catalogue record for this book is available from the British Library.

ISBN 1 85996 208 4

BIOS Scientific Publishers Ltd
9 Newtec Place, Magdalen Road, Oxford OX4 1RE, UK
Tel. +44 (0)1865 726286. Fax +44 (0)1865 246823
World Wide Web home page: http://www.bios.co.uk/

Typeset by Phoenix Photosetting, Chatham, UK
Printed by Cromwell Press Ltd, Trowbridge, UK

Front cover: Co-cultured muscle and nerve cells stained by double immunofluorescence. See Section 9.2.2 and legend on page 44.

Contents

Abbreviations   xi

Preface   xiii

Key to symbols   xv

1.	**Introduction**	**1**

Definition   1
History and development   1
References   3

2.	**Production of Antibodies**	**5**

Immunization   5
Testing   6
Region-specific antibodies   6
Monoclonal antibodies   8
Selection of antibodies by phage display   9
Characteristics of a 'good' antibody   10
References   11

3.	**Preparation of Tissue for Immunocytochemistry**	**13**

Fixation   13
    Cross-linking fixatives   14
    Precipitant fixatives   15
    Combination fixatives   15
Fixed, paraffin-embedded tissue   16
Fresh material – frozen sections and cell preparations   16
    Frozen sections   17
    Whole cell preparations   18
Pre-fixed, non-embedded tissue   20
    Pre-fixed frozen sections   20
    Pre-fixed Vibratome sections   21
    Whole-mounts   21
    Permeabilization   21

Freeze-dried tissue   22
Tissue storage   22
  Paraffin blocks and sections   22
  Frozen blocks and sections   23
  Cell preparations   23
Adherence of sections and cell preparations to slides   24
Antigen retrieval in fixed tissues   24
  Washing   24
  Protease treatment   24
  Heat-mediated antigen retrieval   26
References   29

Colour plate section   **33**

4. Visualizing the End-product of Reaction   **45**

Fluorescent labels   46
  Advantages   46
  Disadvantages   46
  Uses of immunofluorescence   46
  Fluorescein   47
  Rhodamine   47
  Phycoerythrin   48
  AMCA   48
  Other fluorophores   48
  Fluorescent counterstains   48
Enzyme labels   49
  Peroxidase   49
  Alkaline phosphatase   52
  Glucose oxidase   52
  β-D-Galactosidase   53
Gold labels   53
  Colloidal gold   53
  Nanogold   54
Other labels   55
  Biotin   55
  Haptens   55
  Radioisotopes   55
References   56

5. Non-specific Staining due to Tissue Factors   **59**

Causes of non-specific binding   60
  Charged sites   60
  Hydrophobic attraction   60
  Fc receptors   60
Prevention of non-specific binding   60

Other problems   61
  Endogenous enzymes   61
  Endogenous biotin   61
  Autofluorescence   61
References   62

6. Methods   **63**

General considerations   63
  Buffers   63
  Antibody diluent and storage   64
  Antibody dilution relative to reaction time, temperature
    and technique   65
Methods   67
  Nature of antibodies (IgG)   67
  Application of antibodies to preparations   69
  Direct (one-step) method   70
  Indirect (two-step) method   72
  Three-layer methods   73
  Avidin–biotin methods   76
References   79

7. Testing Antibodies: Specificity and Essential Controls   **81**

Testing a new primary antibody   81
  A primary antibody with a known localization   81
  A primary antibody with an unknown localization   85
  Negative control for polyclonal antibodies – normal serum   85
  Negative controls for monoclonal antibodies   85
  Testing for non-specific binding of second and third
    reagents   86
Non-specific or unwanted specific staining due to antibody
  factors   86
  Unwanted specific staining of unknown antigens   86
  Non-specific binding of antisera to basic proteins   86
  Unwanted specific cross-reactivity of anti-immunoglobulins   87
  Cross-reactivity of the primary antibody with related
    antigens   87
Remedies for non-specificity due to tissue factors   89
  Blocking binding sites with normal serum   89
  Absorption with tissue powder   89
Remedies for non-specificity due to heterogeneity
  of the antibody   89
  Dilution   89
  Affinity purification   89
Remedies for non-specificity due to cross-reactivity   90
Essential staining controls   90

Negative controls   90
Positive controls   90
Experimental controls   91
References   91

8. Increasing Sensitivity and Enhancing Standard Methods   93

Increasing sensitivity   93
Immunogold with silver enhancement   93
Build-up methods   94
Tyramine signal amplification (TSA)   97
Intensification of the peroxidase/DAB/H_2O_2 product   100
Post-reaction intensification   100
Intensification during the peroxidase reaction   101
References   101

9. Multiple Immunostaining   103

Double direct immunostaining with separately labelled primary antibodies   103
Double immunostaining with primary antibodies raised in different species, or of different immunoglobulin sub-class   104
Double immunoenzymatic method   104
Double immunofluorescence method   106
Triple immunostaining   106
Unlabelled primary antibodies from the same species   106
The problem   106
Elution methods   107
Indirect double immunostaining without elution   109
References   113

10. Immunocytochemistry for the Transmission Electron Microscope   115

Principles   115
Fixation   115
Pre-embedding immunocytochemistry   116
Non-embedding immunocytochemistry   116
Processing to resin   117
Labels   118
Sectioning resin blocks   119
Pre-treatment   120
Immunolabelling procedure   121
Immunolabelling with peroxidase   121
Amplification   122
Contrasting   122

Multiple labelling   122
References   123

**11. *In Vitro* Methods for Testing Antigen–Antibody
      Reactions   125**

Radioimmunoassay   126
Enzyme-linked immunosorbent assay (ELISA)   126
Western blotting   127
Dot blots   127
References   128

12. Applications of Immunocytochemistry   129

Histopathological diagnosis   129
  Controls   130
  Choice of antibody   131
  Tips for diagnostic laboratories   131
Research   133
Quantification   133
  Confocal microscopy   134
  Flow cytometry and fluorescent antibody cell sorting
    (FACS)   135
  Simpler methods of quantification   135
Non-immunocytochemical uses of labelled probes   136
  Receptor localization   137
  Lectin histochemistry   137
  In situ hybridization of nucleic acids   138
References   138

Appendix: Technical Notes   141

Buffers for diluting antibodies and rinsing   141
  Phosphate-buffered normal saline (PBS)   141
  Tris-buffered normal saline (TBS)   141
Antibody diluent and storage of antibodies   142
  Double dilutions   142
Adherence of preparations to slides   143
  Coating slides with *poly*-L-lysine   143
  Coating slides with silane   144
Blocking endogenous peroxidase reaction   144
  Paraffin sections   144
  Milder methods for cryostat sections and whole-cell
    preparations   145
  Blocking endogenous biotin   146
Enzyme pre-treatment   146
  Trypsin   146

Protease   147
Pepsin   147
Neuraminidase   148
Heat-mediated antigen retrieval using a microwave oven   148
Enzyme development methods   150
Peroxidase   150
Alkaline phosphatase   153
Glucose oxidase   154
β-D-Galactosidase   155
Intensifying the peroxidase/DAB reaction product   156
Following standard development   156
During development   156
Immunostaining methods   158
Initial procedures   158
Immunostaining – all preparations   160
Immunogold staining with silver enhancement   161
Silver acetate auto-metallography   163
Double immunoenzymatic staining   164
Primary antibodies from different species   164
Primary antibodies from the same species, heat-blocking
method   165
Post-embedding electron microscopical
immunocytochemistry using epoxy resin-embedded
tissue and an indirect immunogold method   166
Absorption specificity control (liquid phase)   168
References   169

Index   **171**

Abbreviations

ab	antibody
ABC	avidin-labelled biotin complex
AEC	3-amino-9-ethylcarbazole
ag	antigen
APAAP	alkaline phosphatase–anti-alkaline phosphatase
APES	aminopropyltriethoxy silane
ARK	Animal Research Kit
BSA	bovine serum albumin
CD	cluster of differentiation
CARD	catalysed reporter deposition
CSA	catalysed signal amplification
DAB	diaminobenzidine
DNP	dinitrophenyl aminoproprionitrile
ELISA	enzyme-linked immunosorbent assay
FACS	fluorescent antibody cell sorting
FITC	fluorescein isothiocyanate
IC	immunocytochemistry
Ig	immunoglobulin
IGSS	immunogold staining with silver
IH	immunohistochemistry
LCA	leukocyte common antigen
PAP	peroxidase–anti-peroxidase
PBS	phosphate-buffered saline
PCR	polymerase chain reaction
PEG	polyethylene glycol
PGP	protein gene product
PLP	periodate-lysine-paraformaldehyde
RIA	radioimmunoassay
TBS	Tris-buffered saline
TSA	tyramine signal amplification
UV	ultraviolet

Preface

Immunocytochemistry, the accurate localization of tissue constituents with labelled antibodies, was fathered by A.H. Coons in the 1940s, grew up during the sixties and seventies, and in its maturity has become an indispensable investigative technique in diagnostic histo- and cytopathology and many branches of biomedical science. Its versatility allows it to be used on whole cells or on tissue sections, whether from frozen or fixed and embedded samples, and at both light- and electron-microscopical levels. Immunocytochemistry can be combined with other localization methods such as histological or histochemical staining and *in situ* hybridization of nucleic acids. It is applicable to plant as well as animal tissue, the only requirements being a specific antibody to the antigen in question, suitable preservation of the antigen and a revelation method sensitive enough to depict even low quantities of antigen.

Despite its well-established position, the technique, like so many, has its own tricks of the trade. Newcomers need to know not only how to perform the tests, but also why the various steps are necessary, how to get the best results, what are the essential controls and what to do when things go wrong. These can all be learnt from experienced teachers, but after spending a lot of time teaching a seemingly endless stream of novices in the field, we decided in 1980 to write an introductory text to save ourselves some effort. We used the notes successfully in conjunction with practical courses, and in 1984 they were expanded into the first edition of this book (published by Oxford University Press). A revised edition followed in 1987.

This third edition has been re-organised and updated. There is an expanded section on tissue preparation and increased coverage of cytological preparations. New methods of multiple staining are also included. We hope that it will provide a practically directed basis for carrying out current immunocytochemical methods with enough theoretical information to allow a newcomer to understand the whys and wherefores of the technique. The text is backed up by a reference list for readers wanting further depths of knowledge.

We are grateful to the many colleagues who have helped with suggestions and by providing illustrations.

Julia M. Polak
Susan Van Noorden

Safety

Attention to safety aspects is an integral part of all laboratory procedures, and both the Health and Safety at Work Act and the COSHH regulations impose legal requirements on those persons planning or carrying out such procedures.

In this and other Handbooks every effort has been made to ensure that the recipes, formulae and practical procedures are accurate and safe. However, it remains the responsibility of the reader to ensure that the procedures which are followed are carried out in a safe manner and that all necessary COSHH requirements have been looked up and implemented. Any specific safety instructions relating to items of laboratory equipment must also be followed.

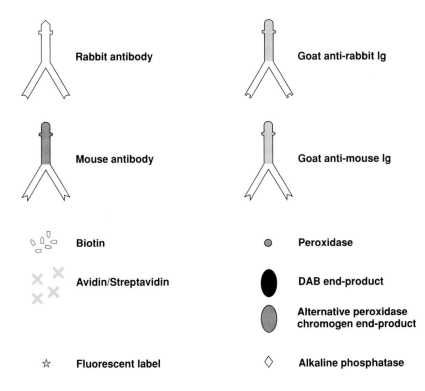

Rabbit antibody

Goat anti-rabbit Ig

Mouse antibody

Goat anti-mouse Ig

Biotin

Peroxidase

Avidin/Streptavidin

DAB end-product

Alternative peroxidase chromogen end-product

Fluorescent label

Alkaline phosphatase

Key to symbols used in diagrammatic figures.

1 Introduction

A distinction is sometimes made between immunohistochemistry (IH) as immunoreactions on tissue sections and immunocytochemistry (IC) as immunoreactions on cells. This terminology has a place in suppliers' catalogues where the application of an antibody is specified as being suitable for cytological use (IC) or use on sections (IH), perhaps with paraffin sections distinguished as IHP, and other acronyms for other applications. However, the principles of all tissue immunoreactions are the same. They apply to cell structures or cell products, whether the cells are contained within their tissue matrix or dissociated, and the term immunocytochemistry will be used throughout this book.

1.1 Definition

There are several uses in biology and medicine for the strong and specific attraction between an antigen and an antibody, including the measurement of antigen in tissue extracts by radioimmunoassay (RIA), analysis of Western blots and the sorting and analysis of populations of dispersed cells after labelling with a fluorescent antibody, i.e. fluorescent antibody cell sorting (FACS). However, immunocytochemistry is the only technique which can identify an antigen in its tissue or cellular location. Thus the definition of immunocytochemistry is the use of labelled antibodies as specific reagents for localization of tissue constituents (antigens) *in situ*.

1.2 History and development

The practice of immunocytochemistry originated with Albert H. Coons and his colleagues (Coons *et al.*, 1941, 1955; Coons and Kaplan, 1950) who were the first to label an antibody with a fluorescent dye and use it

1

to identify an antigen in tissue sections with a fluorescence microscope. As a result of this work much of the uncertainty has now been removed from some aspects of histopathology which were previously entirely dependent on special stains, with interpretation sometimes precariously based on intuition and deduction. Because an antigen–antibody reaction is absolutely specific, positive identification of tissue constituents can now be achieved, though there are still problems as will become apparent in the following pages.

The first fluorescent dye to be attached to an antibody was fluorescein isocyanate, but fluorescein isothiocyanate soon became the label of choice because the molecule was much easier to conjugate to the antibody and more stable (Riggs *et al.*, 1958). Fluorescein compounds emit a bright apple-green fluorescence when excited at a wavelength of 490 nm.

Following the early work, and as better antibodies to more substances became available, the technique has been enormously expanded and developed. New labels have been introduced, including red, yellow, blue and green fluorophores and a variety of enzyme labels which, when developed, can give differently coloured end-products, visible in a conventional light microscope. The methods used to develop the enzyme labels are the standard ones used in histochemistry to identify native enzymes in the tissue. The first enzyme to be used was horse-radish peroxidase (Nakane and Pierce, 1966; Avrameas and Uriel, 1966). Other enzymes include alkaline phosphatase (Mason and Sammons, 1978), glucose oxidase (Suffin *et al.*, 1979) and β-D-galactosidase (Bondi *et al.*, 1982). The end-product of reaction of some of these enzyme reactions can be made electron-dense, but other intrinsically electron-dense labels such as ferritin (Singer and Schick, 1961) have been used for electron microscopical immunolabelling and colloidal gold particles, introduced by Faulk and Taylor (1971), are likely to remain the label of choice for this technique (see Beesley 1993). Antibodies have been labelled with radioactive elements and the immunoreaction visualized by autoradiography, and some other labels in addition to colloidal gold particles, for example latex particles, can be used in scanning electron microscopy.

Among the techniques, the first modification of the original, directly labelled antibody was the introduction of the two-layer indirect method (see Section 6.2). This was followed by the unlabelled antibody–enzyme methods. These avoid entirely chemical conjugation of an enzyme label to an antibody and the consequent reduction of both enzyme and antibody activity. Other methods involved the use of a second antigen (hapten) as an antibody label, visualized by a further antibody raised to the hapten (Cammisuli and Wofsy, 1976; Jasani *et al.*, 1981), the exploitation of the strong attraction between avidin and biotin (Guesdon *et al.*, 1979; Hsu *et al.*, 1981), and numerous ways of improving the specificity and intensity of the final reaction product and of carrying out multiple immunostaining.

Some of these methods are described here, and the Appendix gives details of the basic techniques. However, the subject is too vast to be covered completely in this handbook, which aims only to introduce the concept, and the reader is referred to several publications which provide more detail on selected aspects (Polak and Van Noorden, 1986; Sternberger, 1986; Bullock and Petrusz, 1982, 1983, 1986, 1989; Larsson, 1988; Beesley, 1993; Cuello, 1993; Jasani and Schmid, 1993; Leong, 1993). In addition, several scientific journals provide reviews of anti-bodies and papers on new methods and applications, including: *Applied Immunohistochemistry* (now *Applied Immunohistochemistry and Molecular Morphology*), *Journal of Histochemistry and Cytochemistry, The Histochemical Journal, Histochemistry and Cell Biology* (formerly *Histochemistry*) and *Journal of Cellular Pathology*.

References

Avrameas S, Uriel J. (1966) Méthode de marquage d'antigène et d'anticorps avec des enzymes et son application en immunodiffusion. *C. R. Acad. Sci. Paris Sér. D* **262**, 2543–2545.

Beesley JE. (ed.) (1993) *Immunocytochemistry, A Practical Approach.* Oxford University Press, Oxford.

Bondi A, Chieregatti G, Eusebi V, Fulcheri E, Bussolati G. (1982) The use of β-galactosidase as a tracer in immunohistochemistry. *Histochemistry* **76**, 153–158.

Bullock GR, Petrusz P. (eds) (1982) *Techniques in Immunocytochemistry*, Vol. 1. Academic Press, London.

Bullock GR, Petrusz P. (eds) (1983) *Techniques in Immunocytochemistry*, Vol. 2. Academic Press, London.

Bullock GR, Petrusz P. (eds) (1986) *Techniques in Immunocytochemistry*, Vol. 3. Academic Press, London.

Bullock GR, Petrusz P. (eds) (1989) *Techniques in Immunocytochemistry*, Vol. 4. Academic Press, London.

Cammisuli S, Wofsy L. (1976) Hapten-sandwich labelling, III. Bifunctional reagents for immunospecific labelling of cell surface antigens. *J. Immunol.* **117**, 1695–1704.

Coons AH, Creech HJ, Jones RN. (1941) Immunological properties of an antibody containing a fluorescent group. *Proc. Soc. Exp. Biol. Med.* **47**, 200–202.

Coons AH, Kaplan MH. (1950) Localization of antigen in tissue cells. *J. Exp. Med.* **91**, 1–13.

Coons AH, Leduc EH, Connolly JM. (1955) Studies on antibody production. I: A method for the histochemical demonstration of specific antibody and its application to a study of the hyperimmune rabbit. *J. Exp. Med.* **102**, 49–60.

Cuello AC. (ed.) (1993) *Immunohistochemistry II*, IBRO Handbook Series: Methods in the Neurosciences, Vol. 14. J. Wiley and Sons, Chichester.

Faulk WR, Taylor GM. (1971) An immunocolloid method for the electron microscope. *Immunochemistry* **8**, 1081–1083.

Guesdon JL, Ternynck T, Avrameas S. (1979) The uses of avidin–biotin interaction in immunoenzymatic techniques. *J. Histochem. Cytochem.* **27**, 1131–1139

Hsu SM, Raine L, Fanger H. (1981) Use of avidin–biotin–peroxidase complex (ABC) in immunoperoxidase techniques; a comparison between ABC and unlabeled antibody (PAP) procedures. *J. Histochem. Cytochem.* **29**, 577–580.

Jasani B, Schmid KW. (1993) *Immunocytochemistry in Diagnostic Histopathology.* Churchill Livingstone, Edinburgh.

Jasani B, Wynford Thomas D, Williams ED. (1981) Use of monoclonal anti-hapten antibodies for immunolocalisation of tissue antigens. *J. Clin. Pathol.* **34,** 1000–1002.

Larsson L-I. (1988) *Immunocytochemistry, Theory and Practice.* CRC Press, Boca Raton, Florida.

Leong AS-Y. (1993) *Applied Immunohistochemistry for the Surgical Pathologist.* Edward Arnold, Melbourne.

Mason DY, Sammons RE. (1978) Alkaline phosphatase and peroxidase for double immunoenzymatic labelling of cellular constituents. *J. Clin. Pathol.* **31,** 454–462.

Nakane PK, Pierce Jr GB. (1966) Enzyme-labeled antibodies: preparation and application for the localization of antigens. *J. Histochem. Cytochem.* **14,** 929–931.

Polak JM, Van Noorden S. (eds) (1986) *Immunocytochemistry, Modern Methods and Applications*, 2nd Edn. Butterworth Heinemann, Oxford (originally John Wright and Sons, Bristol).

Riggs JL, Seiwald RJ, Burkhalter JH, Downs CM, Metcalf T. (1958) Isothiocyanate compounds as fluorescent labeling agents for immune serum. *Am. J. Pathol.* **34,** 1081–1097.

Singer SJ, Schick AF. (1961) The properties of specific stains for electron microscopy prepared by conjugation of antibody molecules with ferritin. *J. Biophys. Biochem. Cytol.* **9,** 519–537.

Sternberger LA. (1986) *Immunocytochemistry*, 3rd Edn. John Wiley and Sons, New York.

Suffin SC, Muck KB, Young JC, Lewin K, Porter DD. (1979) Improvement of the glucose oxidase immunoenzyme technique. *Am. J. Clin. Pathol.* **71,** 492–496.

2 Production of Antibodies

This book is not the place for a detailed description of antibody production, but it may be of practical use to know something about the basic procedure in order to understand why antibodies can be so variable.

2.1 Immunization

Antibodies, which are mainly γ-globulins, are raised by immunizing rabbits (or mice, guinea pigs, etc.) with antigen. The antigen must be completely pure or (preferably) synthetic to ensure as specific an antibody as possible. Despite this, the resulting antiserum will not be directed solely to the injected antigen. The antibodies produced by the donor animal will be directed to various parts of the antigen molecule and the antiserum is therefore known as polyclonal. It may also contain antibodies to any carrier protein (see below) and native antibodies that may react with tissue components. A positive-appearing immunoreaction cannot, therefore, be assumed to be due to the specific, desired antigen–antibody reaction unless stringent controls are carried out. It may be necessary to immunize many animals in order to end up with even one usable antiserum, because little is known about what makes an animal react to a foreign protein, and the production of antibodies is still a matter of chance.

If the antigen is large, for example an immunoglobulin, it can be used by itself to immunize an animal. If it is as small as many bioactive peptides, or if the molecule itself is not immunogenic, it must be combined with a larger one for immunization. The small molecule (hapten) is chemically coupled (e.g. by glutaraldehyde or carbodiimide) to a larger 'carrier' protein molecule, such as bovine serum albumin, thyroglobulin or limpet haemocyanin. The larger complex is a better stimulant of antibody formation than the small molecule alone. The donor animal's serum will contain a mixture of antibodies, reactive with diferent amino acid sequences of the hapten and the carrier molecule, but the antibodies to the carrier molecule will either not react with the tissue to be

5

stained (unless it were, for example, limpet tissue and the carrier protein had been limpet haemocyanin) or can be absorbed out, if necessary, by addition of the carrier protein to the antibody before use.

At an appropriate interval after the primary injection (usually subcutaneous) the animal is given a booster injection. Subsequently, blood is taken from it for testing for antibodies. No standard time course can be given for antibody raising, which is a highly individual procedure. The blood is then centrifuged to remove red blood cells. Although the resulting fluid containing the antibody is plasma, not serum, as the fibrin has not been removed, the working solution is illogically known as an antiserum. For further reading on antibody production, see the book by Johnstone and Thorpe (1996).

2.2   Testing

The next step is to test for the presence of antibody. The antiserum may be tested by an enzyme-linked immunosorbent assay (ELISA) against the pure antigen, or by RIA if there is one available, or by a blotting technique (see Chapter 11 for a discussion of these techniques); but by far the most satisfactory way of testing for an antibody to be used in immunocytochemical staining is by immunocytochemistry on known positive tissue. This is because in the *in vitro* methods mentioned above, pure antigen only is offered to the potential antiserum, and thus any interfering, unwanted reactions due to other constituents of the serum are not detected. By immunocytochemistry, the antiserum may be evaluated for the quality of the specific staining against the 'background', and if the background staining is unacceptably high and cannot be eliminated, the antiserum must be abandoned. Another reason for preferring an immunocytochemical test concerns the 'avidity' or 'stickiness' of the antibody (see Section 2.6). A useful antibody for immunocytochemistry must combine strongly with its antigen so that it is not washed off the tissue during the staining procedure.

In RIA, competition takes place between radiolabelled and unlabelled antigen for binding to the antibody, leading to an equilibrium between the two, depending on the proportion of each available. However, a 'good' antibody for RIA is often not good for immunocytochemistry and *vice versa*. The ELISA technique resembles the immunocytochemical technique more closely.

2.3   Region-specific antibodies

The various couplers used to join the carrier protein and the hapten for immunization react preferentially with different functional groups of the hapten molecule. Glutaraldehyde, for example, attaches primarily to amino groups. The free portion of the hapten molecule, distant from the combining site, is most likely to stimulate antibody formation; thus if the only free amino group on the hapten molecule is the NH_2-terminal, then immunization with that hapten after coupling with glutaraldehyde is likely to produce antibodies to the free C-terminal. Carbodiimide, on the other hand, will react with free amino or carboxyl groups; thus immunization with a carbodiimide-coupled hapten, having these groups only at either end of the molecule, is likely to produce a mixture of antibodies directed to the C- and N-terminals. A knowledge of the structure of the hapten is thus essential if the coupler is to be chosen intelligently so that the likely region-specificity of the resulting antibody may be forseen (Szelke, 1983).

It is often advantageous to have an antibody that is specific for only a certain area of the molecule: for instance, in cases where two antigens to be identified have amino acid sequences in common and the antibody is required to distinguish between them (e.g. the intestinal hormones, gastrin and cholecystokinin, which share a C-terminal peptide sequence). Most antibodies will only recognize sequences of four to eight amino acids; however, antigenic sites need not be straight chain sequences of amino acids, but could be merely spatially adjacent sequences, created by folding of the amino acid chain. Attempts to characterize an antibody by assaying or absorbing against fragments of the antigen are not necessarily reliable because there is a danger that a small fragment of an antigen in solution may lose the particular molecular configuration which gave it antigenicity when it was part of a whole molecule. It is usually only by chance that a serum will contain antibodies to the desired part of the molecule, but it is occasionally possible to use synthetic fragments for immunization. With the proviso noted above, the resulting antiserum is then more likely to be specific for those fragments. Unfortunately, the smaller the amino acid sequence used for immunization, the less immunogenic it is, so the chances of obtaining a good antibody to a peptide fragment are slim. Another problem is that the shorter the amino acid sequence, the more likely it is to be common to several different peptides. Thus it may be essential to immunostain with antibodies to several regions of the molecule, for example, the N-terminal, mid-portion and C-terminal amino acid sequences. Results of staining with such antibodies can confirm that the antigen being localized is the genuine molecule or suggest that a related, but not identical, molecule is being identified (Larsson, 1980). Commercially available antibodies should have been tested extensively before being offered for sale but, even with these, caution must be exercised before their absolute specificity can be relied on.

2.4   Monoclonal antibodies

The *in vitro* production of pure antibodies was introduced by Köhler and Milstein (1975). For a clear account of the development of the concept see the work of Milstein (1980).

Antibodies are produced in mice, and activated B-lymphocytes from the spleen, the source of the antibodies, are fused with cultured myeloma (plasma cell tumour) cells from mice of the same strain that are not producing antibody. The fusion results in hybrid cells which continue to grow and divide in culture and also produce antibodies. Unfused cells and non-productive fused cells are eliminated, and the culture of the productive cells is continued. One hybrid cell produces only one type of antibody so the cultured hybrid myeloma ('hybridoma') cells are gradually cloned by limiting dilution (testing the culture medium for secreted antibody at each stage) into cell lines derived from a single cell (monoclonal), in which all the cells produce the same antibody. The testing procedure consists of screening the culture fluid from the various clones for the desired antibody by RIA, ELISA or, preferably, immunocytochemistry, if that is what the antibody is to be used for. Unwanted clones are discarded. For details of monoclonal antibody production and testing see the work of Ritter (1986) and Ritter and Ladyman (1995). As the culture can be stored until further production is required, the method allows for a continuous supply of standard antibodies. The culture fluid contains about 10 µg ml^{-1} of antibody and can be used undiluted or diluted as required by the immunocytochemical method. Alternatively, the hybridoma cells can be grown as an ascitic tumour inside a host mouse. The fluid around the tumour will contain concentrated antibody, about 10 mg ml^{-1}. The disadvantage is that it may also contain proteins secreted by the host animal, and may therefore approach the state of a polyclonal antiserum.

The great advantage of monoclonal antibodies is their absolute specificity for a single sequence or 'epitope' on the antigen molecule. The problems associated with the multiple antibodies contained in polyclonal antisera thus do not arise and immunostained preparations are usually very clean. However, monospecificity for a particular epitope does not necessarily rule out cross-reactivity if the monoclonal antibody happens to be directed to an antigenic sequence shared by more than one substance, whether known or unsuspected, and in this case a monoclonal antibody would offer no advantage over a polyclonal one.

A possible disadvantage of a monoclonal antibody derives from its very monospecificity. A polyclonal antiserum is probably multivalent, consisting of antibodies to several regions of the antigen molecule, providing a strong detecting capacity. A monoclonal antibody, reactive with only one site on the molecule, may result in fewer antibody molecules being bound to the antigen and subsequently detected by the labelling method, resulting in weaker staining. Similarly, the one particular epitope of interest on the antigen molecule may be altered by fixation or processing, so as to be unavailable for reaction, resulting in no staining. This is why some monoclonal antibodies will only react with fresh or frozen material and not with fixed paraffin sections. A polyclonal anti-serum would have more chances of attaching to different epitopes on the molecule that might be less altered. Another possibility for producing antibodies suitable for localization on fixed tissue is to immunize with 'fixed' antigen. In any case, it is essential to screen antibodies in the sys-tem in which they are to be used (see Section 2.2).

In addition to the use of monoclonal antibodies as pure antibodies to known antigens, monoclonal antibodies may be raised to unknown anti-gens and used as markers for particular cell types or cytoplasmic con-stituents. By cloning antibody-making hybridoma cells resulting from the immunization of mice with human thymocytes, several series of monoclonal antibodies to human T lymphocytes were produced and used to separate the cell types immunocytochemically (Kung *et al.*, 1979). The antibodies act against constituents of the cell membranes, but it may be several years after production of the antibodies that the molecular nature of the many cell membrane antigens is discovered (Boyd, 1987). Monoclonal antibodies against cell and tissue components produced by many workers are now periodically revised, and given specific 'clusters of differentiation' (CD) numbers according to the cell or component stained; for example the different sub-types of T and B lymphocytes.

'Good' antibody characteristics (see Section 2.6) apply to monoclonal as much as to polyclonal antibodies, but the availability of such highly specific tools has truly revolutionized immunocytochemistry and its applications in histopathology and cell biology (Gatter *et al.*, 1985).

2.5   Selection of antibodies by phage display

A third way of producing antibodies is to exploit the immense variety of potential antibodies available in B-cells of normal or immunised animals or humans by using a variety of genetic strategies, the most common of which is phage display technology.

Messenger RNA is isolated from B-cells and reverse transcribed into cDNA. This is subjected to the polymerase chain reaction (PCR) with primers specific for the V domains of the IgG molecule that form the extreme end of the antigen-binding fragment (Fv), which is the part of the molecule that actually binds to the antigen (see *Figure 6.1*). The V domains are assembled to form a single chain Fv (scFv) that is then fused to the gene encoding a coat protein of filamentous bacteriophage. When this genetic construct is introduced into the genome of the phage, the antibody protein that it produces is displayed on the surface of the phage. Other antibody fragments such as F(ab) fragments can also be displayed on the surface of phage in the same way. Phage producing the desired antibody can be selected by affinity purification using the appropriate antigen and amplified further. Bacteria (*E. coli*) are infected with the phage and replication of both the bacteria and the phage within them produces a large quantity of antibody that can be isolated and affinity purified. Increasingly, artificial V domains are being made, which are not derived from B cells, thus widening the range of binding sites that can be isolated.

Human antibodies produced in this way can be modified, for example by being tagged with a toxin. The toxic antibody can be used to target a tumour without the danger of the patient producing antibodies against a mouse monoclonal antibody, tagged with a toxin, which might prevent repeated treatment with the same animal-derived antibody.

Phage display antibodies for immunocytochemistry can be labelled with biotin or directly fused with an enzyme such as alkaline phosphatase.

There are many other uses for antibodies and other proteins produced by phage display. For a review and expansion of this very simplified introductory account see George, 1995.

2.6   Characteristics of a 'good' antibody

The main requirement for a good antibody is that it shall be of high affinity for its antigen, in other words that its binding sites fit well with the antigenic sites on its specific antigen and do not attach to other antigens. The avidity, or binding strength, is a connected property depending on the number of fitting sites between the antigen and antibody (Roitt *et al.*, 2001). Immunocytochemistry requires antibodies of high avidity (stickiness), so that they are not washed off the preparations during the staining process.

It is usual to find that the unwanted antibodies in an antiserum are less avid than the antibody to the required antigen and are therefore, fortunately, washed off. However, it should be emphasized that great care must be taken when dealing with histological sections of damaged tissue containing areas of necrosis and with whole cells in smears or unwashed cultures or cytospins, because these preparations seem to attract and hold antibodies non-specifically. Very thorough controls are essential for these, as for all preparations.

The titre or concentration of the antibody is also very important. A high titre allows a high dilution which, in immunocytochemistry, means that in a polyclonal antiserum the population of unwanted antibodies that might react with tissue components is diluted out. High dilutions are also advantageous in that they allow for the fullest use of the available quantity of good antibodies, which are expensive to produce. Unfortunately, the factors leading to antibodies of high avidity and titre are unknown.

If monoclonal antibodies are being used, the dilution factor becomes less important as unwanted reactions are absent and there is, theoretically, an unlimited supply of antibody. However, these reagents are expensive and for reasons of economy, monoclonal antibodies should be diluted to a point where the available antigen is saturated (the serial dilution lower than the one at which the reaction begins to decline; see Section 6.1.3). If the antibody concentration is known, monoclonal antibodies for immunocytochemistry can be diluted to a concentration of between 1 and 10 μg ml^{-1}, at which level they are usually effective.

References

Boyd AW. (1987) Human leukocyte antigens: an update on structure, function and nomenclature. *Pathology* **19**, 329–337.

Gatter KC, Heryet A, Alcock C, Mason DY. (1985) Clinical importance of analysing malignant tumours of uncertain origin with immunohistological techniques. *Lancet* **i**, 1302–1305

George AJT. (1995) Production of antibodies using phage display libraries. In *Monoclonal Antibodies: Production, Engineering and Clinical Application* (eds MA Ritter, HM Ladyman). Cambridge University Press, Cambridge, pp. 142–165.

Johnstone A, Thorpe R. (1996) *Immunochemistry in Practice*, 3rd Edn. Blackwell Scientific Publications, Oxford.

Köhler G, Milstein C. (1975) Continuous cultures of fused cells producing antibodies of predefined specificity. *Nature* **256**, 495–497.

Kung PC, Goldstein G, Reinherz EL, Schlossman SF. (1979) Monoclonal antibodies defining distinctive human T-cell surface antigens. *Science* **206**, 348–349.

Larsson L-I. (1980) Problems and pitfalls in immunocytochemistry of gut peptides. In *Gastrointestinal Hormones* (ed. GB Jerzy-Glass). Raven Press, New York, pp. 53–70.

Milstein C. (1980) Monoclonal antibodies. *Sci. Amer.* **243**, 56–64.

Ritter MA. (1986) Raising and testing monoclonal antibodies for immunocytochemistry. In *Immunocytochemistry, Modern Methods and Applications*, 2nd Edn (eds JM Polak, S Van Noorden). Butterworth–Heinemann, Oxford, pp. 13–25.

Ritter MA, Ladyman H. (eds) (1995) *Monoclonal Antibodies, Production, Engineering and Clinical Application*. Cambridge University Press, Cambridge.

Roitt I, Brostoff J, Male D. (2001) *Immunology*, 6th Edn. Mosby, London.

Szelke M. (1983) Raising antibodies to small peptides. In *Immunocytochemistry, Applications in Pathology and Biology* (eds JM Polak, S Van Noorden). J. Wright and Sons, Bristol, pp. 53–58.

3 Preparation of Tissue for Immunocytochemistry

The essential conditions for immunocytochemistry are summarized in *Table 3.1* Successful immunostaining requires tissue antigens to be made insoluble and yet their antigenic sites must be available to the applied antibody without great alteration of their tertiary structure. In addition, the tissue architecture must be preserved (fixed) so that the immunoreactive cell or organelle may be identified in context. It used to be thought that good tissue fixation meant poor antigen availability, due to the strong cross-linking of tissue proteins by the conventional aldehyde fixatives, but nowadays such fixatives can be used even, in some cases, with osmium tetroxide post-fixation for electron microscopy. This advance is partly the result of generally improved antibodies and more rigorously controlled techniques, but is also dependent on using the correct pre-treatment (see Section 3.8).

Table 3.1: Essential conditions for immunocytochemistry

1. Preservation of the antigen in tissue context (fixation)
2. Specific and sensitive staining with absence of non-specific staining
3. Well characterized antibodies
4. Efficient labelling and detection

3.1   Fixation

The purpose of fixation is to preserve all components of a tissue sample in their true situation, without diffusion. In addition, the tissue must be protected from osmotic damage (swelling or shrinking). For immunocytochemical purposes the antigen to be localized must be made insoluble, yet remain available for reaction with an applied antibody.

Since immunocytochemistry has many uses in diagnostic histopathology (see Chapter 12), an ideal antibody for histological applications will react strongly with its antigen in tissue fixed in formalin, embedded in paraffin and sectioned according to routine procedures, preferably without pre-treatment of the sections. Many such antibodies are now in use. Various resins provide alternative embedding media for electron microscopy and may also be used for thin (1 μm) sections for light microscopy (see Chapter 10).

Some antigens are more readily available in fresh-frozen tissue, but frozen sections are more demanding in treatment than paraffin sections and are inferior in microscopical structure, so paraffin is preferred where possible. Diagnostic immunocytochemistry is of great importance for blood smears and cytological specimens from fluids and fine needle aspirates. These 'whole cell' peparations require specialized treatment.

Choice of a fixative is important for successful immunostaining and it may be worth trying several different types of fixation, following published results and adapting them for the antigen to be localized, taking into consideration the structure of the antigen molecule and the mechanism of action of particular fixatives. If the tissue to be processed contains a large amount of proteolytic enzymes (e.g. gut or pancreas) and the tissue is to be lightly fixed, it may be useful to include a protease inhibitor such as Trasylol, 4000 U ml^{-1} (Bayer) or Bacitracin, 100 μgml^{-1} (Sigma) in the fixative and in any rinsing or storage buffer before freezing or dehydrating the tissue. The structure of the tissue will probably be better and some labile antigens such as the neuropeptide, vasoactive intestinal polypeptide (VIP) will stand a better chance of preservation. It is, in any case, very important to process the tissue as freshly as possible, whatever the type of fixation used.

3.1.1   Cross-linking fixatives

Formalin (aqueous solution of formaldehyde) and glutaraldehyde are cross-linking fixatives – that is, they form links (hydroxy-methylene bridges) between reactive end-groups of adjacent protein chains. Fixed proteins can retain their antigenicity only if the cross-linking does not affect the amino acid sequences that bind to the antibody. Cross-linking fixatives are often essential for fixing the smaller proteins such as bioactive neuropeptides which are, presumably, not long enough to be made insoluble by precipitant fixatives.

In solution, formalin exists in several different forms depending on the pH of the solution, which also influences the effectiveness of formalin solutions as cross-linking fixatives. At low pH there is a preponderance of the more reactive $^+CH_2(OH)$ carbonium ions, while at higher pH, the less reactive $CH_2(OH)_2$ groups predominate. Thus formalin or paraformaldehyde buffered at pH 7 or higher is a milder fixative than unbuffered formalin, and is often recommended as a fixative for immunocytochemistry. Nevertheless, simple formal saline (10% aq. commercial formalin containing 0.9% sodium chloride) is often used routinely and is adequate for many immunocytochemical applications.

Formaldehyde solutions penetrate tissue faster than glutaraldehyde but fix more slowly. Combinations of glutaraldehyde and formaldehyde are sometimes used, particuarly for electron microscopy (see Chapter 10). However, the main consideration regarding fixation for immunocytochemistry is to fix as lightly as possible, compatible with preserving the tissue and making the antigen insoluble. The longer a piece of tissue is left in formalin, the greater will be the degree of cross-linking produced. Some of this can be reversed by washing the tissue thoroughly in water or buffer for several hours before it is processed further, or by some form of pre-treatment of the tissue section (see Section 3.8).

Various other, milder, cross-linking fixatives have been used, in particular for peptide antigens. These include diethyl pyrocarbonate (DEPC), used in vapour form to fix freeze-dried tissue and parabenzoquinone, used both in solution and as a vapour (Pearse and Polak, 1975; Bishop *et al.*, 1978) (see Section 3.5).

3.1.2   Precipitant fixatives

The second major group of fixatives consists of protein precipitants, such as alcohol and acetone. These fixatives denature proteins by destroying the hydrophobic bonds which hold together the tertiary (three-dimensional) conformation of the protein molecule. The primary and secondary structures of the protein are left intact, so the amino acid sequences acting as antigenic sites remain available to their antibodies. Alcohol has been recommended as a fixative for tissue that is to be immunostained after paraffin embedding and sectioning. Battifora and Kopinski (1986) recommended fixing two parallel samples, one in formalin for good structural preservation and one in alcohol for immunocytochemistry. Mixed precipitant fixatives such as Carnoy's fluid (ethanol, chloroform and acetic acid) may also be useful, followed by paraffin embedding. However, modern methods of antigen retrieval from formalin-fixed sections (Section 3.8.3) make this less important.

Fixatives consisting of alcohol (usually methanol), acetone or combinations of these with formalin are often used for whole-cell preparations or for cryostat sections from fresh-frozen material. Precipitant fixatives are useful when the antigen is a large protein, but prefixation in formalin or another cross-linker is essential for small peptides, which are so soluble that they tend to diffuse and disperse during the brief period of thawing while the frozen section is cut and again while it is picked up on the slide.

3.1.3   Combination fixatives

Picric acid is a precipitant which is often added to formalin in combination formulae such as Bouin's and Zamboni's (Stefanini *et al.*, 1967) fixatives. Bouin's is an acidic fixative (70% saturated aq. picric acid, 10%

commercial formalin, 5% acetic acid) and therefore fixes rapidly because of the preponderance of the active $^+CH_2(OH)$ form of formaldehyde, while Zamboni's is buffered at pH 7.4 and fixes more slowly (see Section 3.1.1). Both fixatives are excellent for preserving the antigenicity of small peptides. Bouin's-fixed material is usually processed using paraffin, while tissues fixed in Zamboni's fluid are generally frozen after fixation or processed using resin for electron microscopical immunocyto-chemistry.

Numerous other combination fixatives have been described. One that was designed specifically to fix glycoproteins for electron microscopy is periodate-lysine-paraformaldehyde (PLP) (McLean and Nakane, 1974), and this has been found useful for preservation of the antigenicity of lymphocyte surface antigens, many of which are glycoproteins, in paraffin-embedded tissue (Pollard et al., 1987). It has also been recommended for frozen sections, following an initial brief fixation with the precipitant fixative, acetone (Hall et al., 1987). The mechanism of action of this fixative has been disputed (Hixson et al., 1981). Fixatives containing zinc salts, with or without formalin, also provide good fixation with antigen survival (Dapson, 1993; Beckstead, 1994). Mercuric chloride is another additive to formalin which often produces both good fixation and good immunocytochemistry, but it is no longer in routine use because of its toxicity.

3.2   Fixed, paraffin-embedded tissue

Fixed tissue blocks are processed and sectioned following routine procedures. Sections should be thinner than 5 µm, if possible, to give sharp localization of the immunoreaction product. Sections should be picked up on suitably coated slides (see Section 3.7) and dried overnight at 37–40°C, or for 20 minutes to 1 hour at 60°C. They should not be heated on a hot plate as this can damage immunoreactivity.

3.3   Fresh material – frozen sections and cell preparations

Although thin paraffin sections provide the most easily handled preparations for immunocytochemistry using non-fluorescent antibody labels at the light microscopical level, non-embedded, fresh material may be used in the form of frozen sections or whole-cell preparations (smears, impressions, cytospins or cultures). The most commonly used fixatives for this type of material are alcohol or acetone, although formalin or simply air-drying may also be suitable. The method of preservation will depend on the antigen to be localized.

3.3.1 Frozen sections

Freezing tissue for frozen sections. Snap-frozen tissue blocks are useful for providing thin cryostat sections for localization of antigens that are destroyed by routine processing or where immunofluorescence is to be used. It is important that the tissue is frozen rapidly to prevent the movement of antigens and to provide good structural preservation. Optimal freezing is obtained by plunging the tissue block into a good heat conductor at a very low temperature. Melting *iso*-pentane is a recommended medium. The *iso*-pentane is contained in a plastic beaker and pre-cooled by suspending the beaker in liquid nitrogen until the *iso*-pentane is solid. It is then allowed to thaw at room temperature until there is enough liquid to allow full immersion of the specimen, but still some solid *iso*-pentane in the beaker. Thus, the *iso*-pentane is maintained at its melting point (–160°C), the lowest temperature at which it can be in a liquid form. The tissue block can be orientated before freezing, either on a drop of OCT embedding medium on a cork disc, or contained in a suitable mould surrounded by OCT. It will freeze almost instantly and can then be transferred to a cryostat or, for storage, to a freezer. The *iso*-pentane can be re-frozen and used again, unless there is danger from infectious agents. Freezing tissue gradually by placing it in a freezer or even on solid carbon dioxide will result in tissue disruption due to formation of large ice crystals within the tissue. This effect is particularly marked in muscle. Antigens are well enough preserved, but accurate localization will be impaired. If tissue is frozen in liquid nitrogen, which is certainly cold enough, the block must be agitated during the freezing process to disperse the layer of warm gaseous nitrogen that forms around the tissue and prevents rapid conduction of heat.

Stored at –70°C, both tissue structure and antigens will be well preserved for years. However, dehydration of the tissue will gradually occur, more rapidly at higher temperatures, eventually making it impossible to section the block.

Sectioning frozen tissue. Sections are usually cut in a cryostat maintained at about –20°C. The block is allowed to reach this temperature for sectioning since at lower temperatures there is a risk that the sections will show cracks or chatter. It may sometimes be necessary to section at higher or lower temperatures.

Sections are picked up from the knife blade on warm (room temperature) slides that have been coated with *poly*-L-lysine or other adhesive (Section 3.7) and are dried at room temperature for at least 45 minutes. Sections containing quantities of fat or connective tissue such as skin may benefit from a longer period of drying to ensure adhesion to the slide. For most purposes, there is no point in drying the sections in the cryostat chamber at –20°C. The tissue has already thawed and re-frozen briefly as it is cut (Pearse, 1980) then thawed again as it is picked up on the warm slide. Re-freezing by putting it into the cryostat chamber or fixing it in very cold acetone before it has dried is probably adding insult to injury. The drying (protein denaturation) process seems to be as important as fixation in preserving both immunoreactivity and tissue structure. Sections allowed to dry for several days usually show no impairment of reaction, though this may depend on the antigen to be localized. Dried sections may be stored at –20°C unfixed or after fixation (Section 3.6.2).

Fixation of frozen sections. Thin frozen sections require only brief fixation. Acetone is the most commonly used fixative and 10 minutes in acetone at room temperature will suffice. The acetone should be kept water-free by storage over Molecular Sieve (Type 3A, BDH 54001) or other desiccant and can be re-used several times unless there is any danger of infectious agents in the tissue. In the authors' laboratories, 500 ml of acetone is used for an arbitrarily decided 100 sections before it is discarded. Following fixation, the acetone is allowed to evaporate at room temperature and the sections can then be immunostained or stored.

An alternative precipitant fixative is 100% methanol or ethanol. Some antigens may require a cross-linking fixative. Formal saline, phosphate-buffered formaldehyde, formaldehyde vapour (from liquid formalin) for very soluble antigens, Zamboni's fixative, PLP, acetone followed by PLP (Hall *et al.*, 1987) and even hexazotised pararosaniline (de Jong *et al.*, 1991) have all been used (Sections 3.1.1, 3.1.2, 3.1.3).

After fixation the sections are either dried (for fixatives that evaporate) or rinsed in water or PBS before being immunostained. It is important that once a previously dried frozen section has been exposed to an aqueous solution it should not be allowed to dry up to the end of the immunstaining process. Drying can produce artifacts that impair the immunoreaction and the appearance of the section.

3.3.2   Whole cell preparations

Blood smears, bone marrow smears and cell imprints
Blood smears. Lymphoprep™-separated Buffy coat cells should be used, preferably mixed with some of the erythrocytes to lend support to the smear and disperse the white cells. The layer of cells should be as thin as possible.

All types of preparation. Smears or imprints are usually air-dried before being fixed and stained for morphology. If they are well dried, it may not be necessary to spread them on poly-L-lysine-coated or other adherent slides; however, this may be a safer procedure to prevent cells floating off the slide during immunostaining. The dried sample can be fixed and treated in the same way as unfixed cryostat sections. Short fixation (90 seconds) in a mixture of acetone and methanol (1:1) is useful for immunocytochemistry on blood smears. Other fixatives such as acetone or alcohol alone can also be used. Acetone provides sharper immunocytochemical localization, but alcohol is better for morphology. Alternative fixation, after air-drying, particularly where a smear has not been washed, is 0.1% formal saline for 2–14 hours, or 10% buffered formalin for 10 minutes, followed by post-fixation for 10 minutes in 100% ethanol (Leong *et al.*, 1999). The weak formalin fixation rehydrates the dried cells and lyses the red blood cells, diminishing background from endogenous peroxidase-like enzymes (catalases). However, after this fixation procedure, heat-mediated antigen retrieval is necessary to achieve full immunoreactivity (Section 3.8.3). The morphological appearance of the cells is not damaged by this treatment.

Cytological preparations. Cells in fluids or from fine needle aspirates are usually washed to remove mucin and debris, then made into cytospin preparations. Where there is adequate cellular material it is useful to make a cell pellet which is fixed in formalin and embedded in paraffin so that many different immunostains can be done on the sections.

A convenient way of treating cytospins is to immerse them immediately in 95%–100% alcohol. The preparations can be kept in this until they are immunostained (this is the usual practice in the authors' routine diagnostic laboratories) or they can be removed, dried and stored in a freezer as for cryostat sections (Section 3.6.2).

Leong *et al.* (1999) suggest rehydration of the dried preparations before staining by immersion for less than 5 minutes in 0.9% sodium chloride (normal saline).

A recently advocated method for cell preparations is liquid-based fixation. The cells are dispersed in the fixative fluid before being cytocentrifuged. The fixative is alcohol-based and contains a mucolytic agent. The morphological appearance of the cells is very good, but for immunocytochemistry of many antigens heat-mediated antigen retrieval is required (Section 3.8.3).

Some antigens are very much better immunostained after formalin than after alcohol fixation even if the extra step of antigen retrieval is necessary. Oestrogen and progesterone receptors are an example (Suthipintawong *et al.*, 1997).

Cultured cells. Immunocytochemistry can be performed on cells grown in dishes, wells or flasks, but the volume of reagent required to cover these preparations makes the cost prohibitive and it is difficult to view the stained preparation on an inverted microscope. Therefore, it is preferable that adherent cells are grown on coverslips or multiwell slides. Non-adherent cells can be stained using cytospin preparations or cell smears.

Consideration must be given to the material of the supporting coverslip or slide. Some plastics become opaque in acetone and other solvents so fixation in acetone would be impossible, as would dehydration and clearing at the end of the reaction. Glass or special plastic coverslips should be used if acetone fixation is required. Some materials are fluorescent, so if immunofluorescence is to be used, this must be taken into account. Multi-chamber glass or plastic slides are a useful base for cell cultures as the individual chambers can be immunostained with different antibodies and control procedures, and the slide can be viewed in a fluorescence or optical microscope as required. Cells grown on a coverslip or slide should not be thicker than a monolayer as overlapping cells could prevent accurate visualization of the stained antigen.

The medium should be washed off in buffer to remove proteins before the cells are fixed. The type of fixation depends, as always, on the antigen with the added complication that cell surface antigens may have to be distinguished from cytoplasmic ones. If the cellular localization of the antigen is unknown, it is a good idea to apply the primary antibody to separate preparations, one fixed before and the other after the application. Antigens on the cell surface will be able to bind to their antibody in the unfixed cells, but if the antigen is cytoplasmic, the antibody will not be able to penetrate the cell membrane until the lipid components of the membrane have been modified by fixation.

Acetone and alcohol are fixatives that will break down the lipids of the membranes while fixing cell proteins. Formalin-based fixatives usually prevent penetration of the cell membranes by antibodies so fixation must be followed by an additional permeabilization step. This can be achieved by soaking the cell preparations in buffer containing a detergent before immunostaining. Saponin (0.1%) has been recommended as less damaging than other detergents to intracytoplasmic membranes (Goldenthal *et al.*, 1985) but Triton X-100 (0.2%) or Tween 20 (0.05–0.2%) may be adequate.

Following formalin fixation, it may be necessary to enhance antigen availability by heat-mediated retrieval (see Section 3.8.3). This does not seem to have adverse effects on cells.

3.4   Pre-fixed, non-embedded tissue

3.4.1   Pre-fixed frozen sections

Formalin-fixed tissue blocks can also be frozen for cryostat sections. Cross-linking fixation is necessary for neuropeptides and this procedure is often useful for obtaining fairly thick sections (20–40 µm) for tracing neuropeptide-containing nerve fibres by immunofluorescence. Before freezing, formalin-fixed blocks must be cryoprotected (e.g. with 13% sucrose) to modify the water content of the tissue (increased by the aqueous fixative) and prevent ice crystal formation during freezing. Background fluorescence may be a problem, but the passage of a nerve may be more easily seen with epi-fluorescence in a thick section than in a thin paraffin one (*Plate 1(b)*). However, laser scanning confocal microscopy, if available, would provide a better way of imaging structures in a thick preparation (Section 12.3.1 and *Plate 16*).

3.4.2   Pre-fixed Vibratome sections

Pre-fixed tissue can also be cut at room temperature on a Vibratome (vibrating knife microtome), avoiding the freezing process. Such sections are necessarily thick and are useful for the same purposes as fixed frozen sections.

3.4.3   Whole-mounts

Whole-mounts such as membranes (e.g. diaphragm, iris) can also be fixed and stretched onto a coated slide before immunostaining. Whole mount preparations may require permeabilization before the imunostaining procedure (Costa *et al.*, 1980).

3.4.4   Permeabilization

Pre-fixed, non-embedded material has not been subjected to solvents during processing , and it may be necessary to break down the lipids of the cell membranes in these thick preparations before applying immunoreagents. It may be helpful to soak the preparations in buffer containing a detergent (Triton X-100, 0.2%, or Tween 20, 0.05–0.2%) before beginning the immunostaining procedure, and/or to dilute the reagents in a detergent-containing buffer. Larsson (1981) recommends a detergent soak procedure for paraffin and resin sections as well as for fixed frozen sections, but in our experience this is not usually necessary; however, addition of detergent to the rinsing buffer during the immunostaining procedure may help to prevent non-specific attachment of protein to the sections (see Section 5.2).

An alternative method is to dehydrate the preparations through graded alcohols to xylene and then to rehydrate them before staining (Costa *et al.*, 1980). The disadvantage of this method is that solvent-labile antigens may be leached out.

3.5   Freeze-dried tissue

Antigens that are particularly soluble or liable to diffuse within the tissue may be captured at the moment of snap-freezing the tissue by subsequent drying and vapour fixation.

Freeze-drying consists of snap-freezing a small tissue sample then removing water from it by evaporation under vacuum in a tissue freeze-dryer while it is still frozen. Thus the tissue constituents are preserved *in situ* and denatured. Freeze-drying is usually followed by a 1–3 h exposure at 60°C to fixative vapour, for example generated from paraformaldehyde powder or from parabenzoquinone dissolved in a solvent such as toluene. The dry, fixed tissue is then infiltrated with paraffin under vacuum and sections are cut in the normal way. This method is particularly useful for small peptide antigens (Pearse and Polak, 1975) but it requires special apparatus and the morphological preservation is never good enough for electron microscopy. Improvements in antibodies and method sensitivity have meant that routinely fixed tissue has become suitable for most immunocytochemical reactions and freeze-drying is not much employed now.

3.6   Tissue storage

3.6.1   Paraffin blocks and sections

Antigens seem to survive indefinitely in paraffin blocks so that freshly cut sections are unaffected. There may be problems in paraffin sections stored at room temperature – less so if they are stored at 4°C and even less at −80°C (Grabau *et al.*, 1998; Biddolph and Jones, 1999). However, many antigens can be largely retrieved by heat treatment (Section 3.8.3). Although van den Broek and van de Vijver (2000) reported no antigen retrieval effect on their selected antigens in stored sections, their retrieval method used citrate buffer and only 10 minutes' microwaving. After microwaving in Tris/EDTA buffer, pH 9.0, either for 25 or 35 minutes at the standard temperature (Grabau *et al.*, 1998) or for a longer period at a lower temperature for fragile tissues (Biddolph and Jones, 1999), antigens were well immunostained. Results vary with the antigen stained and the time and temperature of storage, but the best approach, if archival sections are to be stained, seems to be to microwave in Tris/EDTA buffer, pH 9, for at least 25 minutes (see Appendix A.6). Pressure cooker or autoclave times might be different. For sections that were not mounted on suitably coated slides, overnight treatment at 80°C is a safeguard against detachment.

If tinctorially stained sections only are available from the archive, it is often possible to de-stain them, retrieve the antigens and re-stain with an immunocytochemical method (Milios and Leong, 1995; Biddolph and Jones, 1999).

3.6.2   Frozen blocks and sections

Once tissue samples have been frozen, they may be stored at a low temperature (–40 to –70°C) for many months, provided that drying is prevented by exclusion of air from around the tissue as far as possible. This is because dried tissue is very difficult to section in a cryostat, although antigens will still be preserved. Sections from fresh-frozen tissue blocks can be stored after air-drying, or after fixation in a non-aqueous medium such as acetone, by wrapping the slides individually in aluminium foil or plastic 'cling-film' and sealing the collection in a plastic bag or box containing silica gel or other desiccant. The container is put in a deep-freeze (–20°C is adequate). The entire container is allowed to come to room temperature before the slides are removed and unwrapped, so that any water of condensation in the container will be absorbed by the desiccant and will not come into contact with the sections to dissolve soluble antigens. This is particularly important if the sections are not fixed before storage.

Sections from pre-fixed frozen blocks and acetone- or alcohol-fixed cell culture preparations can be stored in the same way after drying on the slides or, for a shorter time, at 4°C. Cytospin preparations or cultured cells that have been fixed in alcohol can also be kept in that solution at room temperature or 4°C for at least some days, and possibly indefinitely, without destroying fixed antigens.

3.6.3   Cell preparations

Formalin-fixed cell-culture preparations can be stored in buffer at 4°C, and alcohol-fixed ones in alcohol until they are used.

For transport of alcohol-fixed cytological preparations or long storage at room temperature, it is advisable to protect antigenic reactivity by coating the cells with a mixture of 10% polyethylene glycol (PEG), MW 1450, in 50% methanol (Maxwell *et al.*, 1999). It may be necessary to heat both the solid PEG and the methanol solution to 60°C before mixing. Prior to immunostaining, the PEG must be removed with methanol.

Cytocentrifuged cells can be stored at –20°C, before or after alcohol or acetone fixation, dried, wrapped and desiccated in the same way as frozen sections. If formalin-fixed, they can be washed in distilled water (to remove salts that might precipitate on the cells) and dried before being stored at –20°C.

It is possible to destain Papanicolaou-stained preparations and re-stain them with an immunocytochemical method, as with paraffin sections (Abendroth and Dabbs, 1995).

3.7 Adherence of sections and cell preparations to slides

Paraffin sections for immunostaining should never be 'baked' on a hot-plate, as this degrades the antigenicity of many substances. They should be thoroughly dried for several hours or overnight at 37°C and, if necessary, heated in a 60°C oven for 10–15 min.

For the simpler techniques, sections will adhere well to clean slides with no coating, particularly if they have been well dried; but to ensure that sections and cells will not detach from slides during the longer procedures of immunocytochemical staining, an adhesive is usually applied to the slide before the section is picked up or the cells are allowed to settle. This is an esssential precaution if antigen retrieval methods have to be used (see Section 3.8).

Among the many slide coatings that have been suggested, a universally applicable one is *poly*-L-lysine (Huang *et al.*, 1983). This imparts a negative charge to the slide which attracts positively charged tissue elements. When high temperature heat-mediated antigen retrieval (Section 3.8.3) is to be used, it is essential to use slides with a stronger charge or adhesive power. Commercially available ones are, for example, Vectabond (Vector Laboratories) or Polysine (Merck) slides, and there are many others. A suitable home-produced alternative is to use slides dipped in aminopropyltriethoxy silane (APES). Slide coating methods are decribed in the Appendix (Section A.3).

3.8 Antigen retrieval in fixed tissues

The process now known as antigen retrieval is applied to aldehyde-fixed tissues in which antigenicity has been reduced by the formation of hydroxy-methylene bridges between components of the amino acid chains of proteins. In many instances, immunoreactivity can be restored without compromising the structure of the tissues.

3.8.1 Washing

The simplest form of reversing the effects of formalin is to wash the tissue well before processing, but this is not usually possible in histopathology laboratories, where rapid turnover of specimens is required.

3.8.2 Protease treatment

A little understood but practical way of revealing strongly cross-linked proteins is to treat the (de-waxed, hydrated) sections with a protease such as trypsin or pronase before immunostaining (*Figure 3.1*) (Huang *et al.*, 1976). It is thought that the enzyme breaks the cross-linking bonds of the fixative with the protein to reveal antigenic sites. A few antigens may be adversely affected and there is also a possibility that large protein molecules (e.g. precursors of bioactive peptides) may be cleaved by the enzyme to smaller molecules (Ravazzola and Orci, 1980). If these peptide molecules display antigenic sites that were not available on the precursor, a false positive reaction for the peptide may occur. In practice, this rarely happens, and protease pretreatment should be tried routinely whenever preliminary immunostaining is found to be inadequate (Curran and Gregory, 1977; Mepham *et al.*, 1979; Finlay and Petrusz, 1982). However, heat-mediated retrieval methods, if effective, are less capricious (Section 3.8.3).

Several different enzymes have been advocated and it is a question of trial or individual preference as to which is used. We use routinely a crude and relatively inexpensive form of trypsin from porcine pancreas which contains some chymotrypsin. In fact, chymotrypsin may be the

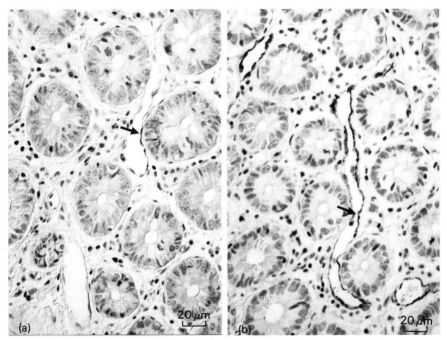

Figure 3.1: Human duodenum immunostained for Factor-VIII-related protein, a marker for endothelial cells, by the PAP method using a rabbit antibody to Factor VIII followed by unconjugated goat anti-rabbit IgG, then rabbit PAP complex. The section shown in (a) was immunostained without pretreatment; the adjacent section in (b) was immunostained after treatment at 37°C with 0.1% trypsin in 0.1% calcium chloride at pH 7.8. Note the intensification of the reaction in the vessel walls in (b) (arrow). Formalin-fixed, paraffin-embedded section; haematoxylin counterstain. (Reproduced from Polak, J.M. and Van Noorden, S. (1997) *An Introduction to Immunocytochemistry* 2nd edition, Microscopy Handbook 37, BIOS Scientific Publishers (Figure 3.1, p. 17).

active ingredient, since purified trypsin gave poor results, and chymo-trypsin alone can be used (Brozman, 1980). A few antigens are prefer-entially revealed by other enzymes, for example IgE by protease XXIV. This enzyme is also preferred for displaying membrane-bound immunoglobulins and complement in renal and skin autoimmune dis-eases (Howie *et al.*, 1990). The enzymes must be used at their optimal pH for a time determined by experiment in the laboratory. Methods for several are given in the Appendix (Section A.5).

The time for which enzyme treatment is carried out may be critical within 1 or 2 min if optimal revelation is to be combined with acceptable tissue structure. Over-digestion can lead to destruction of the tissue matrix. It is important that everyone in the laboratory uses a standard method, so that 'a 10 min treatment with trypsin' is understood to mean, for instance, that the slides are (or are not) warmed to 37°C before treat-ment and that the enzyme solution has been freshly prepared in a pre-warmed (or cold) solvent. Results are more likely to be uniform when a rack of slides is immersed in a large volume, but when expensive enzymes such as protease XXIV have to be used, it is more economical simply to cover the preparations with drops of solution, staggering the application times to give the correct incubation period for each.

The term 'correct' is really only applicable when tissues of uniform size have been fixed for a standard time. The longer a tissue sample is left in formalin, the greater will be the degree of cross-linking and the longer the digestion time needed to reveal a given antigen. Different proteins require different digestion times, even after standard fixation, and different batches of enzyme may vary in effectiveness; so it is necesary for each laboratory to establish the treatment time required by each antigen, to be prepared to try a range of digestion times for samples received from other laboratories, and to test each new batch of enzyme received from the supplier if its composition is different from that of the previous batch. We have found also that the same antigen in different locations may require different digestion times. For instance, immunoglobulin deposits in basement membranes of skin or renal glomeruli are optimally revealed only after 35–45 minutes in protease XXIV, compared with 10 min for the same molecules in plasma cells. This may be partly due to the increased visibility of the immunostained deposits after the plasma in renal glomerular loops has been digested away. This process can be monitored microscopically (Howie *et al.*, 1990).

3.8.3   Heat-mediated antigen retrieval

A major step forward in immunocytochemistry was made with the dis-covery that some antigens previously unreactive in formalin-fixed, paraffin-embedded tissue, even after protease treatment, could be 'retrieved' by heating sections in a solution of a heavy metal salt in a microwave oven without deleterious effects on the structure of the tissue (Shi *et al.*, 1991). The experiment was undertaken to try to improve immunocytochemical reactions because it had previously been shown that both microwave treatment on dry sections (Sharma *et al.*, 1990) and metal ions as a post-formalin fixation protein precipitant (Abbondanzo *et al.*, 1991) had an enhancing effect. Subsequently, it was shown that the rather toxic heavy metal salts could be replaced by simple buffers such as citrate buffer at pH 6.0 (Cattoretti *et al.*, 1993). It was shown that heating, rather than microwaving, is important in the retrieval process, since boiling the sections in a pressure cooker (Norton *et al.*, 1994) or autoclaving them (Bankfalvi *et al.*, 1994) in the buffer solution achieved the same effect. The pH of the buffer may also play a part in antigen retrieval (Shi *et al.*, 1993).

A possible explanation of the process has been put forward by Morgan *et al.* (1994), who suggested that heating provides the energy not only to rupture the hydroxyl bonds formed by the fixative with the protein antigen, freeing some antigens, but also to release tissue-bound calcium ions which contribute to tighter bonds with the fixative. They showed that the salt solutions in which the sections are heated are, in fact, all able to chelate or precipitate calcium to varying degrees and thus remove released calcium from the sections, breaking fixative bonds permanently and revealing antigens. The most effective solutions (EDTA, EGTA) were also the best calcium chelators. However, this hypothesis is probably not of universal application (Shi *et al.*, 1999). Even precipitant-fixed or lightly formalin-fixed fresh tissue material may respond to heat-mediated antigen retrieval. Miller *et al.* (2000) report improved immunostaining results on air-dried smears after pressure-cooking. Alcohol-fixed smears were not affected. Sodium dodecyl sulphate (SDS) was effective as a denaturing agent in revealing antigens on formalin-fixed frozen sections and cultured cells (Brown *et al.*, 1996) and 0.2 M boric acid, pH 7.0 was better than Tris/EDTA pH 9.0 for low temperature retrieval of antigens in archival paraffin sections of lymphoid tissue (Biddolph and Jones, 1999).

Examples of diagnostically important antigens that are revealed by this method are the oestrogen and progesterone receptor antigens in breast carcinomas (*Plate 2(a)*, p. 33) and the CD 30 antigen in Reed–Sternberg cells in Hodgkin's lymphoma. Published lists of antigens and their preferred retrieval methods abound and are of some help in deciding on the 'correct' method to use (Cuevas *et al.*, 1994; Werner *et al.*, 1996); but again, it is probably necessary for each laboratory to optimize its antigen retrieval method for each antigen–antibody combination, as separate epitopes on the same antigen may respond differently, so that identification with antibodies other than the ones cited may require different conditions.

One of the factors that has been well established is that with a standard temperature the period of heating required to reveal antigens is proportional to the period for which the tissue was fixed in formalin.

This is, presumably, because the amount of cross-linking increases progressively with time of fixation. If the heating temperature is raised, the time of exposure may be reduced. Heating at a relatively low temperature (80–90°C) overnight can be as effective as 2 minutes in a pressure cooker or 10 minutes in a microwave oven at 800 watts. This may be useful when sections are fragile or are not mounted on suitably coated slides (Peston and Shousha, 1998; Biddolph and Jones, 1999; Koopal *et al.*, 1998).

It seems that antigens are more tolerant of heat-mediated than of enzyme-mediated retrieval methods, in that antigens and tissues survive heating longer than the minimum time required to reveal the antigen. However, some antigens require a longer period of heating than others, the time usually falling between 2 and 30 min. Some laboratories give all their sections a standard time, equating to the longest required, which could help to overcome vagaries of variable fixation time, while others prefer to treat each antigen individually. As with the enzyme digestion methods, it is important for standard procedures to be set up. The method used in our laboratories is given in the Appendix (Section A.6). It is of great importance that sections are firmly attached to slides for methods involving heating in buffer solutions (see Section 3.7).

Heat-mediated antigen retrieval increases the sensitivity of immunoreactions to such an extent that, for antigens on which it confers improved immunoreactivity, it is usually necessary to dilute a primary antibody considerably further than for the standard, non-retrieval method. Retrieval sometimes also reveals unwanted background reactivity, which can be a nuisance and require further treatment to block it (see Chapters 5 and 7).

The ultimate goal of a universal retrieval solution and heating method may be within sight. Tris/EDTA buffer, pH 9 (Appendix A.6) is suitable for most antigens and may increase the sensitivity of the reaction so much that the primary antibody may be diluted further, even beyond the dilution used with heating in citrate buffer (K. Miller, personal communication). Proprietary antigen retrieval solutions from commercial companies can also be very useful. Nevertheless, although very many antigens respond to heat-mediated methods and can be used as markers in histopathology, some are still preferentially revealed by protease treatment (e.g. CD 21). Others may be even more enhanced by heat treatment followed by a short exposure (seconds) to a protease (e.g. LP34 in our hands) or protease pre-treatment followed by heat treatment (Frost *et al.*, 2000). Some are unaffected by formalin fixation and are quite adequately demonstrated without any pre-treatment (e.g. in our hands, proliferating cell nuclear antigen, PCNA, with antibody PC10). It is even possible that heat pre-treatment can destroy the immunoreactivity of some antigens. Antigen retrieval methods have been reviewed extensively by Taylor *et al.*, (1996), Shi *et al.*, (1997) and Miller *et al.*, (2000).

References

Abbondanzo SL, Allred DC, Lampkin S, Banks PM. (1991) Enhancement of immunoreactivity among lymphoid malignant neoplasms in paraffin-embedded tissues by refixation in zinc sulfate-formalin. *Arch. Pathol. Lab. Med.* **115**, 31–33.

Abendroth CS, Dabbs DJ. (1995) Immunocytochemical staining of unstained versus previously stained cytologic preparations. *Acta Cytol.* **39**, 379–386.

Bankfalvi A, Navabi H, Bier B, Bocker W, Jasani B, Schmid KW. (1994) Wet autoclave pretreatment for antigen retrieval in diagnostic immunohistochemistry. *J. Pathol.* **174**, 223–228.

Battifora H, Kopinski M. (1986) The influence of protease digestion and duration of fixation on the immunostaining of keratins. A comparison of formalin and ethanol fixation. *J. Histochem. Cytochem.* **34**, 1095–1100.

Beckstead JH. (1994) A simple technique for preservation of fixation-sensitive antigens in paraffin-embedded tissues. *J. Histochem. Cytochem.* **42**, 1127–1134.

Biddolph SC, Jones, M. (1999) Low-temperature, heat-mediated antigen retrieval (LHTMAR) on archival lymphoid sections. *Appl. Immunohistochem. Mol. Morphol.* **7**, 289–293.

Bishop, AE, Polak JM, Bloom SR, Pearse AGE. (1978) A new universal technique for the immunocytochemical localisation of peptidergic innervation. *J. Endocrinol.* **77**, 25P–26P.

Brown D, Lydon J, McLaughlin M, Stuart-Tilley A, Tyszkowski R, Alper S. (1996) Antigen retrieval in cryostat tissue sections and cultured cells by treatment with sodium dodecyl sulfate (SDS). *J. Histochem. Cytochem.* **105**, 261–267.

Brozman M. (1980) Antigenicity restoration of formaldehyde-treated material with chymotrypsin. *Acta Histochem.* **67**, 80–85.

Cattoretti G, Pileri S, Parravicini C *et al.* (1993) Antigen unmasking on formalin-fixed, paraffin-embedded tissue sections. *J. Pathol.* **171**, 83–98.

Costa M, Buffa R, Furness JB, Solcia E. (1980) Immunohistochemical localization of polypeptide in peripheral autonomic nerves using whole mount preparations. *Histochemistry* **3**, 157–165.

Cuevas EC, Bateman AC, Wilking BS, Johnson PA, Williams JH, Lee AHS, Jones DB, Wright DH. (1994) Microwave antigen retrieval in immunocytochemistry: a study of 80 antibodies. *J. Clin. Pathol.* **47**, 448–452.

Curran RC, Gregory J. (1977) The unmasking of antigens in paraffin sections of tissue by trypsin. *Experientia* **33**, 1400–1401.

Dapson RW. (1993) Fixation for the 1990s: a review of needs and accomplishments. *Biotechnic Histochem.* **68**, 75–82.

De Jong JP, Voerman JS, Leenen PJM, van der Sluis-Gelling A, Ploemacher RE. (1991) Improved fixation of frozen lympho-haemopoietic tissue sections with hexazotized pararosaniline. *Histochem. J.* **23**, 392–401.

Finlay JCW, Petrusz P. (1982) The use of proteolytic enzymes for improved localization of tissue antigens with immunocytochemistry. In *Techniques in Immunocytochemistry,* Vol. 1. (eds GR Bullock, P Petrusz). Academic Press, New York, pp. 239–249.

Frost A, Sparks D, Grizzle WE. (2000) Methods of antigen recovery vary in their usefulness in unmasking specific antigens in immunohistochemistry. *Appl. Immunohistochem. Mol. Morphol.* **8**, 236–243.

Goldenthal KL, Hedman K, Chen JW, August JT, Willingham M. (1985) Postfixation detergent treatment for immunofluorescence suppresses localization of some integral membrane proteins. *J. Histochem. Cytochem.* **33**, 813–820.

Grabau DA, Nielsen O, Hansen S, Nielsen MM, Laenkholm A-V, Knoop, A, Pfeifer P. (1998) Influence of storage temperature and high-temperature antigen retrieval buffers on results of immunohistochemical staining in sections stored for long periods. *Appl. Immunohistochem.* **6**, 209–213.

Hall PA, Stearn PM, Butler MG, D'Ardenne AJ. (1987) Acetone/periodate-lysine-paraformaldehyde (PLP) fixation and improved morphology of cryostat sections for immunohistochemistry. *Histopathology* **11**, 93–101.

Hixson DC, Yep JM, Glenney Jr JR, Walborg Jr EF. (1981) Evaluation of periodate/lysine/paraformaldehyde fixation as a method for cross-linking plasma membrane glycoproteins. *J. Histochem. Cytochem.* **29**, 561–566.

Howie AJ, Gregory J, Thompson RA, Adkins, MA, Niblett AJ. (1990) Technical improvements in the immunoperoxidase study of renal biopsy specimens. *J. Clin. Pathol.* **43**, 257–259.

Huang S, Minassian H, More JD. (1976) Application of immunofluorescent staining in paraffin sections improved by trypsin digestion. *Lab. Invest.* **35**, 383–391.

Huang WM, Gibson S, Facer P, Gu J, Polak JM. (1983) Improved section adhesion for immunocytochemistry using high molecular weight polymers of L-lysine as a slide coating. *Histochemistry* **77**, 275–279.

Koopal SA, Iglesias Coma M, Tiebosch ATMG, Suurmeijer AJH. (1998) Low-temperature heating overnight in Tris-HCl buffer pH 9 is a good alternative for antigen retrieval in formalin-fixed paraffin-embedded tissue. *Appl. Immunohistochem.* **6**, 228–233.

Larsson L-I. (1981) Peptide immunocytochemistry. *Prog. Histochem. Cytochem.* **13**

Leong AS-Y, Suthipintawong C, Vinyuvat S. (1999) Immunostaining of cytologic preparations: a review of technical problems. *Appl. Immunohistochem. Mol. Morphol.* **7**, 214–220.

Maxwell P, Patterson AH, Jamieson J, Miller K, Anderson N. (1999) Use of alcohol-fixed cytospins protected by 10% polyethylene glycol in immunocytology external quality assurance. *J. Clin. Pathol.* **52**, 141–144.

McLean IW, Nakane PK. (1974) Periodate-lysine-paraformaldehyde fixative. A new fixative for immunoelectron microscopy. *J. Histochem. Cytochem.* **22**, 1077–1083.

Mepham BL, Frater W, Mitchell BS. (1979) The use of proteolytic enzymes to improve immunoglobulin staining by the PAP technique. *Histochem. J.* **11**, 345–357.

Milios J, Leong AS-Y. (1995) The detection of tissue antigens in previously stained histological sections. *Stain Technol.* **62**, 411–416.

Miller RT, Swanson P, Wick MR. (2000) Fixation and epitope retrieval in diagnostic immunohistochemistry: a concise review with practical considerations. *Appl. Immunohistochem. Mol. Morphol.* .**8**, 228–235.

Morgan JM, Navabi H, Schmid KW, Jasani B. (1994) Possible role of tissue-bound calcium ions in citrate-mediated high-temperature antigen retrieval. *J. Pathol.* **174**, 301–307.

Norton AJ, Jordan S, Yeomans P. (1994) Brief, high-temperature heat denaturation (pressure cooking): a simple and effective method of antigen retrieval for routinely processed tissues. *J. Pathol.* **173**, 371–379.

Pearse AGE. (1980) *Histochemistry, Theoretical and Applied* 4th edition, Volume 1. Churchill Livingstone, Edinburgh, pp. 25–27.

Pearse AGE, Polak JM. (1975) Bifunctional reagents as vapour and liquid phase fixatives for immunohistochemistry. *Histochem. J.* **7**, 179–186.

Peston D, Shousha S. (1998) Low temperature heat mediated antigen retrieval for demonstration of oestrogen and progesterone receptors in formalin-fixed paraffin sections. *J. Cell Pathol.* **3**, 91–97.

Pollard K, Lunny D, Holgate CS, Jackson P, Bird CC. (1987) Fixation, processing and immunochemical reagent effects on preservation of T-lymphocyte surface membrane antigens in paraffin-embedded tissue. *J. Histochem. Cytochem.* **35**, 1329–1338.

Ravazzola M, Orci L. (1980) Transformation of glicentin-containing L-cells into glucagon-containing cells by enzymatic digestion. *Diabetes* **29**, 156–158.

Sharma, HM, Kauffman EM, McGaughy VR. (1990) Improved immunoperoxidase staining using microwave slide drying. *Lab. Med.* **21**, 658–660.

Shi S-R, Key ME, Kalra KL. (1991) Antigen retrieval in formalin-fixed, paraffin-embedded tissues: an enhancement method for immunocytochemical staining based on microwave oven heating of tissue sections. *J. Histochem. Cytochem.* **39**, 741–748.

Shi S-R, Imam A, Young L, Cote RJ, Taylor CR. (1993) Antigen retrieval immunohistochemistry under the influence of pH using monoclonal antibodies. *J. Histochem. Cytochem.* **43**, 193–201.

Shi S-R, Cote RJ, Taylor J. (1997) Antigen retrieval immunohistochemistry: past, present, and future. *J. Histochem. Cytochem.* **45**, 327–343.

Shi S-R, Cote RJ, Hawes D, Thu S, Shi Y, Young L, Taylor CR. (1999) Calcium-induced modification of protein conformation demonstrated by immunohistochemistry: what is the signal? *J. Histochem. Cytochem.* **47**, 463–469.

Stefanini M, De Martino C, Zamboni L. (1967) Fixation of ejaculated spermatozoa for electron microscopy. *Nature (London)* **216**, 173–174.

Suthipintawong C, Leong S-Y, Chan KW. (1997) Immunostaining of estrogen receptor, progesterone receptor, MIB1 antigen and c-erbB-2 oncoprotein in cytologic specimens: a simplified method with formalin fixation. *Diagn. Cytopathol.* **17**, 127–133.

Taylor CR, Shi S-R, Cote, RJ. (1996) Antigen retrieval for immunohistochemistry: status and need for greater standardization. *Appl. Immunohistochem.* **4**, 144–166.

van den Broek LJCM, van der Vijver MJ. (2000) Assessment of problems in diagnostic and research immunohistochemistry associated with epitope instability in stored paraffin sections. *Appl. Immunohistochem. Mol. Morphol.* **8**, 316–321.

Werner M, von Wasielewski R, Komminoth P. (1996) Antigen retrieval, signal amplification and intensification in immunohistochemistry. *Histochem. Cell Biol.* **105**, 253–260.

(a)

(b)

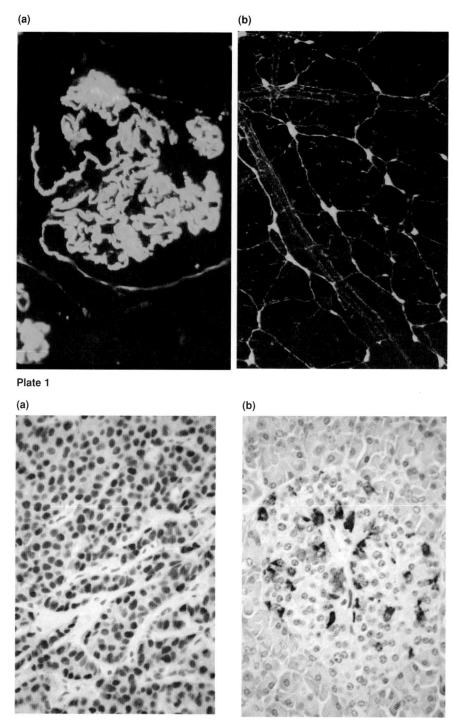

Plate 1

(a)

(b)

Plate 2

See pages 41–44 for legends.

(a)　(b)

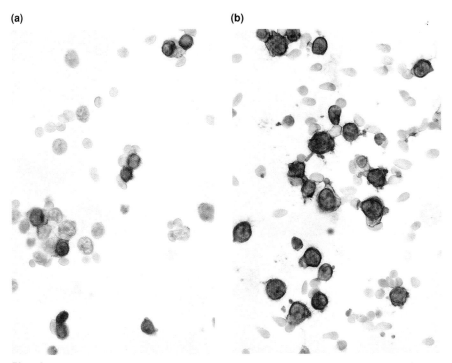

Plate 3

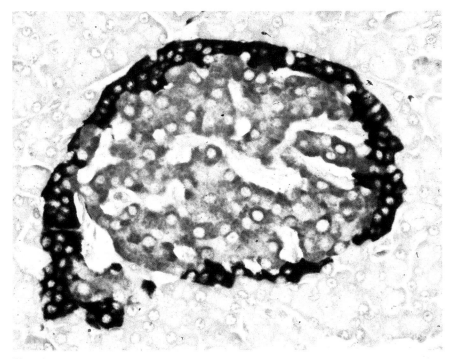

Plate 4

(a)

(b)

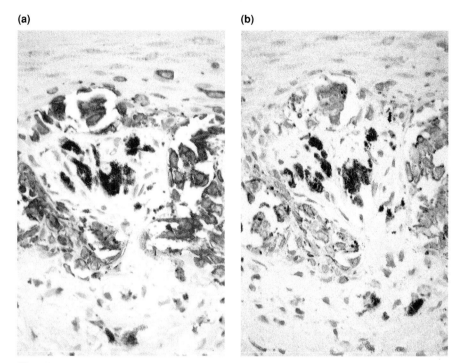

Plate 5

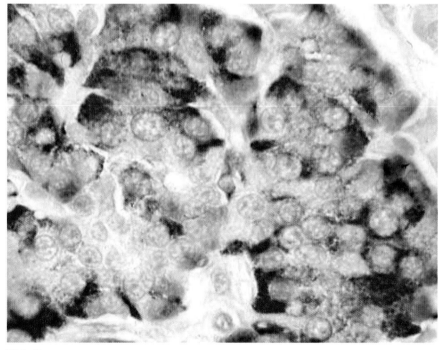

Plate 6

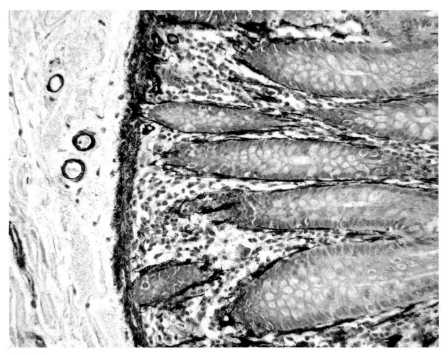

Plate 7

(a) 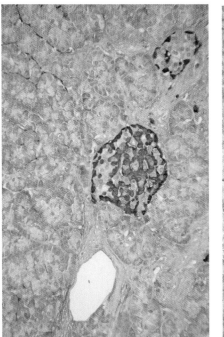   (b)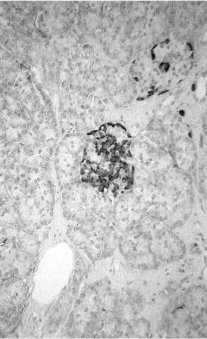

Plate 8

(a)

(b)

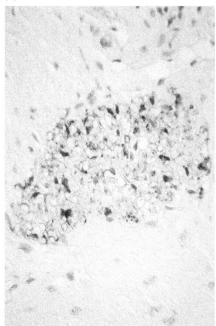

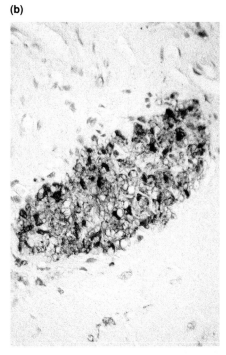

Plate 9

(a)

(b)

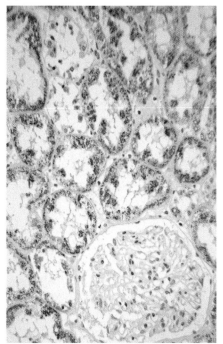

Plate 10

(a)

(b)

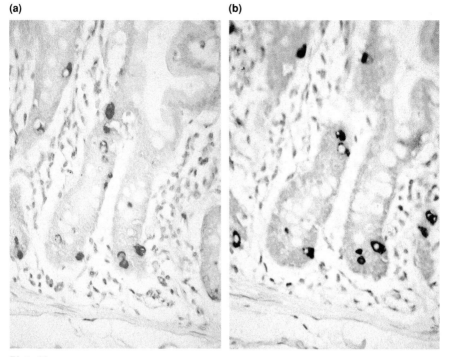

Plate 11

(a)

(b)

(c)

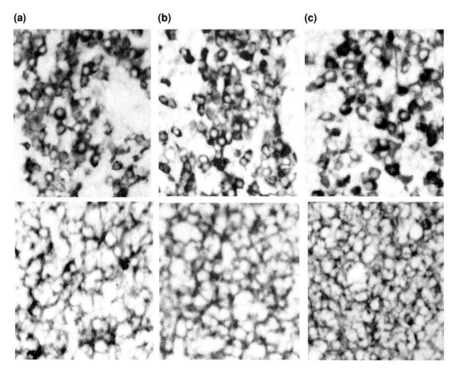

Plate 12

(a)

(b)

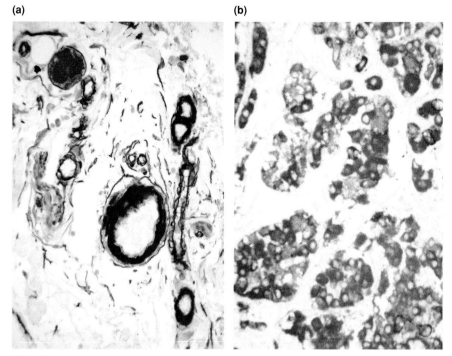

Plate 13

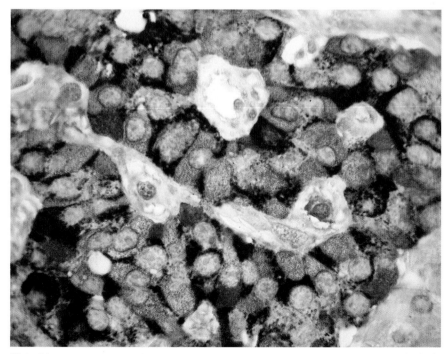

Plate 14

(a)

(b)

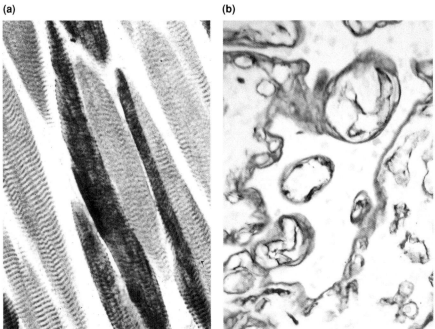

Plate 15

(a)

(b)

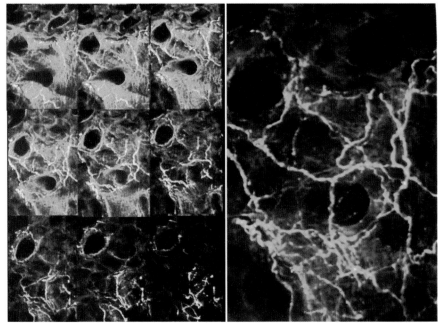

Plate 16

Plate 1: (a) Glomerulus from a human kidney biopsy. Fresh frozen cryostat section (5 µm) post-fixed in acetone, immunostained by a direct method with FITC-conjugated rabbit anti-human IgG. This is a simple and rapid diagnostic method. The result could be available 1 hour after taking the tissue. The patient's circulating auto-antibodies to the glomerular basement membrane become bound to the membrane and can be seen here in a typical linear pattern of distribution. Courtesy of Dr. E. M. Thompson. (b) Rat ileum fixed in Zamboni's fixative. Frozen section (40 µm) of sub-mucous plexus immunostained for protein gene product (PGP 9.5, a general marker for nerves) by indirect immunofluorescence with a rabbit polyclonal antibody to PGP 9.5 followed by FITC-conjugated goat anti-rabbit immunoglobulin. Note the nerve cell bodies and the fine nerve fibres surrounding the blood vessels. Before application of the primary antibody, the section was dehydrated through alcohols to xylene, then rehydrated (Costa *et al.*, 1980).

Plate 2: (a) Human breast carcinoma. Oestrogen receptors immunostained by a peroxidase-labelled streptavidin method. The substrate was H_2O_2 and the chromogen, DAB. Labelling is present in the nuclei of the tumour cells. Formalin-fixed paraffin section pre-treated by microwave heating for 20 min in 0.01 M citrate buffer, pH 6.0; haematoxylin counterstain. (b) Human pancreas. Somatostatin in islet D cells. Immunoperoxidase stain using a PAP method with rabbit anti-somatostatin as the primary antibody. The substrate was H_2O_2 and the chromogen, DAB. Formalin-fixed paraffin section, no pretreatment, haematoxylin counterstain.

Plate 3: Cytospin preparation of cells in lung effusion from a patient with a high grade B-cell lymphoma. The cytospins were fixed in alcohol and immunostained by a peroxidase-labelled streptavidin method using monoclonal primary antibodies. The T and B cell markers illustrated are present on the cell membrane. Haematoxylin counterstain. (a) Normal T-lymphocytes immunostained for CD3 are present among the cells. (b) Many abnormal B-lymphocytes are present, immunostained for CD20.

Plate 4: Rat pancreatic islet. Bouin's-fixed paraffin section immunostained for glucagon with a PAP method. The chromogen is DAB but the end-product of reaction has been made black by incorporation of a nickel salt (Appendix, Section A.8). The section was then immunostained for insulin with an indirect immunoperoxidase method using amino-ethyl carbazole as the chromogen to give a red stain (Appendix, Section A.7.1). Provided that the two antigens are in different sites, the intense black colour of the DAB–nickel reaction product masks any cross-over of antibodies and reaction product from the second reaction, allowing a double stain with two polyclonal antibodies. Haematoxylin counterstain.

Plate 5: Human skin melanoma. Formalin-fixed paraffin section pre treated by microwaving for 20 min in 0.01 M citrate buffer, pH 6.0. Haematoxylin counterstain. (a) Immunostained for Melan A (a melanoma marker) using an alkaline phosphatase-labelled streptavidin method. The substrate was naphthol AS-MX phosphate and the chromogen, Fast Red TR, giving a red reaction product which contrasts with the natural brown colour of the melanin pigment that is present in this section. (b) Negative control. PBS was substituted for the primary antibody. The brown melanin granules can be clearly seen. If an immunoperoxidase stain had been used with DAB as the chromogen it would have been difficult to distinguish the brown immunostain from the brown melanin.

Plate 6: Human pancreatic islet. Simultaneous double immunoenzymatic stain (Appendix, Section A.11.1) immunostained for glucagon with a polyclonal rabbit antibody to glucagon and a mouse monoclonal antibody to insulin followed by biotin-labelled swine anti-rabbit Ig and unconjugated goat anti-mouse Ig. The third layer was β-D-galactosidase-labelled avidin and mouse PAP. This illustrates the turquoise blue reaction product of β-D-glactosidase development with an indigogenic method (Appendix, Section A.7.4) and the red colour from peroxidase development with aminoethyl carbazole as chromogen (Appendix, Section A.7.1). Formalin-fixed paraffin section, no pretreatment, haematoxylin counterstain.

Plate 7: Human colon immunostained for smooth muscle actin by the immunogold method with silver enhancement. Smooth muscle borders submucosal blood vessels and is present in the muscularis mucosae and around the mucosal glands. This is a very intense stain, easily visible even at low magnification. Formalin-fixed paraffin section, no pretreatment, haematoxylin counterstain.

Plate 8: Human pancreas showing immunoperoxidase stain (ABC method) for glucagon in an islet. Formalin fixed paraffin sections, no pretreatment, haematoxylin counterstain. This illustrates reduction in non-specific background staining by dilution of the primary antibody. (a) The primary antibody was used at a dilution of 1/500. (b) The primary antibody was used at a dilution of 1/6000.

Plate 9: This illustrates the increase in sensitivity of an avidin–biotin method compared with a PAP method. Human colon, immunostained for S-100, which can be used as a marker for the axon sheaths of nerves. A nerve ganglion in the myenteric plexus is shown. Formalin-fixed paraffin sections, no pretreatment, haematoxylin counterstain. The primary rabbit antibody was applied to each section for 1 hour at room temperature at the same dilution (1/8000). (a) Subsequent development by the PAP method. (b) Subsequent development by a peroxidase-labelled streptavidin method. The stain is much stronger in (b) than in (a).

Plate 10: Human kidney, formalin-fixed paraffin sections, microwave heated for 10 minutes in 0.01 M citrate buffer, pH 6.0. Haematoxylin counterstain. The section (b) was treated with unlabelled avidin followed by unlabelled biotin to block endogenous biotin (Appendix, Section A.4.3). Peroxidase-labelled streptavidin was applied to both sections. (a) Endogenous biotin is strongly stained and could be confused with a concomitant immunostain using an avidin–biotin method. (b) Endogenous biotin has been completely blocked.

Plate 11: Human duodenum, immunostained for the endocrine cell marker, chromogranin, by an indirect immunoperoxidase method with H_2O_2 and DAB as the chromogen. This illustrates how a weak DAB-peroxidase reaction product can be intensified by post-reaction exposure to copper sulphate (Appendix, Section A.8.1). Formalin-fixed paraffin sections, no pretreatment, haematoxylin counterstain. (a) No post-reaction treatment. (b) After peroxidase development the section was immersed in copper sulphate solution for 2 minutes. The reaction product is considerably darker than in (a).

Plate 12: Human tonsil, formalin-fixed paraffin sections, microwave-heated for 10 minutes in 0.01 M citrate buffer followed by treatment for 10 seconds with 0.1% trypsin solution at pH 7.8. The sections were immunostained by the double immunoenzymatic method to show two distinct antigen localizations (Appendix, Section A.11.1). Where the two antigens are present in the same structure, colour mixing will be evident.

The primary antibody layer was rabbit polyclonal anti-kappa immunoglobulin light chains simultaneously applied with mouse monoclonal anti-lambda light chains. The second layer was a mixture of swine anti-rabbit Ig and biotinylated goat anti-mouse Ig. The third layer was a mixture of rabbit PAP and alkaline phosphatase-labelled streptavidin. Alkaline phosphatase was developed in red (Fast Red TR as chromogen) or blue (Fast Blue BB as chromogen). Peroxidase was the developed brown with DAB as chromogen, black with DAB including a nickel salt in the incubating solution and red with aminoethyl carbazole as chromogen. No counterstain was applied. Plasma cells (upper panel) contain only one type of immunoglobulin and will therefore be positive for either kappa or lambda light immunoglobulin chains, but not both and will show pure colours. In the germinal centres of the tonsil (lower panel) immunoglobulin of many types is present between the cells. The intercellular immunoglobulin will therefore contain both kappa and lambda light chains and will show colour mixing.

(a) Kappa light chain Ig (polyclonal antibody) is shown in brown (peroxidase-DAB) and lambda light chain Ig in blue (alkaline phosphatase-Fast Blue BB). Upper panel, plasma cells; lower panel, intercellular Ig in germinal centre. (b) Kappa light chain Ig (polyclonal antibody) is shown in red (peroxidase-AEC) and lambda light chain Ig in blue (alkaline phosphatase-Fast Blue BB). Upper panel, plasma cells; lower panel, intercellular Ig in germinal centre. (c) Kappa light chain Ig (polyclonal antibody) is shown in black (peroxidase-DAB-Ni) and lambda light chain Ig in red (alkaline phosphatase-Fast Red TR). Upper panel, plasma cells; lower panel, intercellular Ig in germinal centre.

Plate 13: Triple immunoenzymatic immunostaining by two different methods, both including the use of two or more antibodies from the same species. In both cases the three antigens are known to be in different locations. Both sections are from formalin-fixed paraffin embedded tissue. The primary antibody was applied without heat or protease pretreatment. No nuclear counterstain was used. (a) Human colon sub-mucosa. Smooth muscle actin, black; endothelium, blue; nerves, red. Smooth muscle actin was immunostained first with monoclonal primary mouse antibody and mouse PAP. Peroxidase was developed black with DAB as the chromogen and nickel enhancement. Then after blocking any remaining peroxidase, a simulta-neous double stain was done for endothelial cells with mouse monoclonal anti-CD34 and for nerves with rabbit polyclonal anti-PGP 9.5. The second layer was a mixture of biotinylated goat anti-mouse Ig and unconjugated swine anti-rabbit Ig. The third layer was a mixture of alkaline phosphatase-labelled streptavidin and rabbit PAP. Alkaline phosphatase was developed blue with Fast Blue BB as the chromogen and peroxidase red with AEC as the chromogen. The intense black colour of the DAB-nickel reaction product masks any cross-over of mouse anti-bodies from the second reaction. (b) Human anterior pituitary gland, *post mortem* tissue. ACTH, blue; prolactin, red; growth hormone, brown. Three rabbit polyclonal antibodies were used with microwave heat blocking of potentially cross-reacting residual Ig and alkaline phosphatase after each development (Appendix, Section A.11.2). Biotin blocking was also done between the reactions.

Rabbit anti-ACTH, followed by biotinylated goat anti-mouse Ig and alkaline phosphatase-labelled streptavidin. The enzyme was developed blue. After inter-layer blockings, rabbit anti-prolactin was applied and similarly treated, but the alkaline phosphatase was developed red. After repeat microwave blocking (biotin blocking not necessary), rabbit anti-growth hormone was applied, followed by unconjugated swine anti-rabbit Ig and rabbit PAP. Peroxidase was developed light brown with H_2O_2 as substrate and DAB as chromogen.

Plate 14: Human pancreatic islet, formalin-fixed paraffin section, no pretreatment. Quadruple immunostain by a combination of methods. The four hormones stained are known to be present in separate cells so no colour mixing is expected. Pancreatic polypeptide, red; somatostatin, black; insulin, blue; glucagon, brown.

First primary antibody: polyclonal rabbit anti-pancreatic polypeptide, developed with an alkaline phosphatase-labelled streptavidin method. The alkaline phosphatase was developed red with Fast Red TR as the chromogen.

The section was then microwaved to block spare immunoglobulin binding sites and residual alkaline phosphatase activity. Unoccupied avidin and biotin sites were also blocked. The second primary antibody was mouse monoclonal anti-insulin, followed by biotinylated goat anti-mouse Ig, then alkaline phosphatase-labelled streptavidin. The alkaline phosphatase was developed blue.

The blocking steps were repeated and a third primary antibody applied. This was polyclonal rabbit anti-somatostatin, followed by biotinylated swine anti-rabbit Ig and peroxidase-labelled streptavidin, the peroxidase developed black with DAB and a nickel salt as chromogen.

The blocking steps were repeated, including a residual peroxidase block with H_2O_2 but not the biotin block. The final primary antibody was polyclonal rabbit anti-glucagon, followed by unconjugated swine anti-rabbit Ig and rabbit PAP. The peroxidase was developed light brown with DAB.

Plate 15: Double immunostaining with two mouse monoclonal antibodies using the Dako ARK kit. Formalin-fixed paraffin section. No nuclear counterstain. (a) Human skeletal muscle. Fast myosin, brown: slow myosin, blue. No pretreatment. The first primary antibody was monoclonal mouse anti-fast myosin followed by Dako Envision peroxidase-labelled goat anti-mouse Ig. Unoccupied goat anti-mouse Ig binding sites were blocked with mouse Ig (normal mouse serum), then the second primary antibody was applied. This was ARK-biotinylated, monoclonal mouse anti-slow myosin, which was followed by alkaline phosphatase-labelled streptavidin. The alkaline phosphatase was developed blue and the peroxidase brown with H_2O_2 as substrate and DAB as chromogen. (b) Human placenta. Epithelial cytokeratin, red; endothelial CD34, blue. The section was pre-treated with trypsin. The first primary antibody was CD34, followed by unconjugated goat anti-mouse Ig and APAAP. After blocking any unoccupied goat anti-mouse Ig binding sites with normal mouse serum, ARK-biotinylated, monoclonal mouse anti-cytokeratin was applied, followed by peroxidase-labelled streptavidin. The alkaline phosphatase was developed blue and the peroxidase red with AEC as chromogen.

Plate 16: Confocal microscopy (Section 12.3.1). (a) Images captured at nine levels through a thick (100 μm) frozen section of rat lung, pre-fixed in Zamboni's fluid and immunostained by an indirect method for PGP 9.5, a general marker for nerves. The green colour of the fluorescence (FITC label) is artificially shown. There is some background autofluorescence. (b) A combina-tion of the nine images shown in (a) to define the entire area occupied by immunofluorescent nerves in this section.

Front cover: Co-cultured muscle and nerve cells immunostained by a double immunofluorescence method using a mixture of mouse monoclonal anti-actin and rabbit polyclonal anti-PGP 9.5 followed by fluorescein-labelled goat anti-mouse Ig and rhodamine-labelled goat anti-rabbit Ig. Photographs of the green and red fluorescence in the same field were taken separately with appropriate filters and the images combined.

Colour plates kindly sponsored by Dako Ltd.

4 Visualizing the End-product of Reaction

In order for an immunocytochemical reaction to be seen in the microscope, a component of the reaction must carry a label. The first label to be attached to an antibody was a coloured dye (Marrack, 1934) but the resulting intensity was too low for visualization. Nearly all labels that have been used subsequently require additional steps to enhance them to the point of visibility. Fluorescent compounds, the first practical labels, require excitation with light of a specific wavelength to make them emit visible light. Enzymes must react with a substrate and chromogen to produce a visible deposit. Radioactive labels require autoradiographic development. Biotin must itself be labelled or reacted with labelled avidin. Colloidal gold provides an electron-dense label for electron microscopical immunocytochemistry and can be seen in the light microscope if enough is applied, but it is more visible after enhancement with metallic silver.

The methods of attaching labels to antibodies are beyond the scope of this book and will be discussed only briefly. For details, see the work of Sternberger (1986) and Johnstone and Thorpe (1996). The ratio of label to antibody is important, since overloading the antibody with label will reduce its immunoreactivity, but it is also important to ensure that all available antibody is labelled optimally, for maximum efficiency.

Labelling an antibody with fluorescein isothiocyanate (FITC) is done quite simply. Briefly, the method consists of:

1. isolation of IgG from the antiserum;
2. reaction of the IgG with FITC in the right proportions at alkaline pH;
3. separation of antibody from unconjugated FITC on a Sephadex G25 column; and
4. separation of labelled from unlabelled IgG on a DEAE cellulose column.

Labelling with an enzyme requires an additional large molecule such as glutaraldehyde to cross-link the enzyme to the antibody, unless an antigen–antibody reaction is used, as in the preparation of the peroxidase–anti-peroxidase (PAP) complex (Sternberger, 1986). A radioactive label (e.g. $Na^{125}I$) may be conjugated to an antibody via another molecule (Johnstone and Thorpe, 1996) or incorporated within a monoclonal antibody during its production, thus avoiding the conjugation process (Cuello *et al.*, 1982). Antibodies are easily labelled with biotin and kits for doing this are available commercially. Antibodies are attached to colloidal gold particles by non-covalent adsorption (Beesley, 1989).

4.1   Fluorescent labels

4.1.1   Advantages

Fluorescence provides an instantly visible label with excellent contrast when seen against a dark, non-fluorescent background. It is usually used on frozen sections or fresh whole-cell preparations because formalin-fixed tissue tends to be autofluorescent or even, if catecholamines are present, to emit formaldehyde-induced specific fluorescence of a colour approaching that of fluorescein compounds (Falck *et al.*, 1962; Hökfelt and Ljungdhal, 1972). In an enzyme-labelled preparation, frozen sections show structural imperfections, but with immunofluorescence these can be ignored, as the background tissue should only be visible enough to set the specifically labelled structures in context.

4.1.2   Disadvantages

A special microscope is necessary, preferably with epi-illumination, so that neither exciting nor emitted light is lost by passing through the specimen. Different filter sets are required for the different fluorescent markers to prevent transmission of extraneous light. Preparations are not permanent because fluorescent labels are not resistant to dehydration and solvent-based mountants, so aqueous mountants must be used. In addition, fluorescence tends to fade, particularly under exposure to the excitation light; photography is therefore difficult, although mountants incorporating fluorescence preservers are now available. These are recommended, particularly those that harden so that the danger of damaging the preparation by moving the coverslip is removed (e.g. Permafluor from Immunotech). Fluorescent dyes other than fluorescein have a longer life (see below).

4.1.3   Uses of immunofluorescence

In histopathology, immunofluorescence is used as a rapid way of analysing immunoglobulin deposits in frozen sections of renal glomerular membranes and basement membrane of skin. Paraffin sections present some problems in this field, particularly the need for variable protease digestion times depending on the fixation of the specimen, and the presence of endogenous biotin in kidney samples. However, the pattern of deposition, which is diagnostic of different types of disease, is easily seen in immunofluorescent, acetone-fixed cryostat sections (*Plate 1(a)*, see p. 33) (Evans, 1986; Chu, 1986). Identification of circulating auto-antibodies is another common use; the patient's serum is applied to frozen sections of normal tissue and bound antibodies are detected by fluorescein-labelled anti-human immunoglobulins (Scherbaum *et al.*, 1986a,b).

Research applications include visualization of neuropeptides in nerves, when the use of thick frozen sections from a tissue block prefixed in parabenzoquinone, Zamboni's fluid or paraformaldehyde allows the undulating course of a nerve to be seen continuously rather than in discontinuous portions, as would be seen in a thin paraffin section (*Plate 1(b)*) (Bishop *et al.*, 1978).

Immunofluorescence, again used in relatively thick preparations, provides an excellent method for quantification of an antigen or examination of its three-dimensional distribution with a laser scanning confocal microscope (see Section 12.3.1).

Fluorescent antibodies are particularly useful for labelling living cells as they do not kill the cells. They are much used in cell sorting and flow cytometry (see Section 12.3.2).

4.1.4   Fluorescein

The first usable immunocytochemical method employed fluorescein isocyanate as a label for immunoglobulin (Coons *et al.*, 1941, 1955). Later (Riggs *et al.*, 1958) the isocyanate was replaced by the isothiocyanate (FITC), which is a more stable compound. FITC is still used widely as a fluorescent marker, emitting a bright apple-green fluorescence (λ 520 nm) at an excitation wavelength of 495 nm.

4.1.5   Rhodamine

Rhodamine derivatives, such as tetrarhodamine isothiocyanate (TRITC), and Texas Red, which fades less rapidly than TRITC (Titus *et al.*, 1982), fluoresce red at an excitation maximum of 530 nm. Whether red or green fluorescence is used is a matter of individual choice, but a combination of two antibodies, one labelled with a fluorescein and the other with a rhodamine fluorophore, provides a useful method of localizing two antigens in a single preparation. The microscope must have narrow band excitation filters which are used alternately, so that the fluorescein-labelled and rhodamine-labelled antigens are seen only when light of the correct wavelength reaches the preparation. The two colours can be seen together on a double exposure photograph (see front cover image). Digital photography and image capture make the process of superimposing images much easier now, but many microscope companies can supply filters that allow red and green fluorescence to be seen at the same time.

4.1.6 Phycoerythrin

Phycoerythrin, a fluorophore from seaweed (Oi *et al.*, 1982), is another rhodamine derivative, but fluoresces in the same excitation range as fluorescein. It emits a rather weak orange light, but may be useful for double immunofluorescence with fluorescein without the need for changing the filter. This method will only be satisfactory if the two antigens to be localized are present in different structures. If they are co-localized, the stronger green fluorescence of the fluorescein marker will dominate the orange.

4.1.7 AMCA

Blue fluorescence can be provided by 7-amino-4-methyl-coumarin-3-acetic acid (AMCA) (Khalfan *et al.*, 1986), and is another useful label for multiple immunofluorescence labelling (Wessendorf *et al.*, 1990).

4.1.8 Other fluorophores

Several photo-stable, fluorescent compounds have now been introduced for immunofluorescence, developed to overcome the disadvantages of fading and thus even to allow for some quantification of fluorescence. Among them are Alexa dyes such as Oregon Green™ (from Molecular Probes Inc. via Cambridge Bioscience, UK), which has the same spectral properties as fluorescein, but is much brighter and more stable under the excitation beam. There are others in the series as well as the various cyanine-based dyes (CyDyes™, Amersham International, UK), which fluoresce brightly within narrow wavelength limits across the spectrum. The molluscan green fluorescent protein is another useful fluorophore. These fluorescent substances can also be used to label probes for *in situ* hybridization of nucleic acids.

4.1.9 Fluorescent counterstains

A fluorescent tissue counterstain such as Pontamine Sky Blue (Cowen *et al.*,1985) which fluoresces red at the excitation wavelength for fluorescein, can provide a useful background to specific green immunofluorescence. Methyl green fluoresces red and can be used as a nuclear counterstain (Schenk and Churukian, 1974). Red nuclear fluorescence is also provided by propidium iodide and blue fluorescence by DAPI or the Hoechst dye. The periodic acid–Schiff (PAS) method can also provide a red background on suitable tissue. These stains are best performed before the immunostain.

4.2   Enzyme labels

The instability of immunofluorescence prompted the development of a more permanent type of preparation using enzymes as labels. Horseradish peroxidase was introduced in 1966 independently by Avrameas and Uriel and by Nakane and Pierce. Alkaline phosphatase was pioneered by Avrameas (1972) and used by Mason and Sammons (1978), glucose oxidase by Suffin *et al.* (1979), and β-D-galactosidase by Bondi *et al.* (1982). With their specific 'histochemical' substrates and a variety of capturing chromogens, enzyme markers can be developed to give coloured end-products of reaction, usually brown, blue or red, that are easily visible in the ordinary light microscope. Optimal dilutions of antibodies and time of incubation must be based on a standard development time for each enzyme that allows for a full reaction – i.e. leaving the preparation longer in the development solution will not further intensify the reaction product. Any endogenous enzyme of the same type as the label must be inhibited (blocked) so that it cannot react with the substrate and become confused with the applied label.

Immunoenzyme methods can be used on any type of tissue preparation and provide a permanent result. Some of the end-products are soluble in alcohol or clearing agents so a water-based mountant is required. This was formerly a disadvantage for microscopical study and photography because of the refractile nature of the mountants and also because of their propensity eventually to dry back. However, mountants have now been produced that have the same refractive index as glass. After application to the preparation, they dry hard and the preparation can then be secondarily mounted with a synthetic mountant and coverslip for permanence with no loss of resolution (e.g. Aquaperm, Thermo-Shandon, Life Sciences, UK).

Each enzyme has its own advantages and disadvantages, but peroxidase and alkaline phosphatase remain the most popular.

4.2.1   Peroxidase

The pH optimum for peroxidase activity is about 5, but at this pH reaction products tend to be coarse, so histochemical development is usually carried out at pH 6.0–7.6 to slow the reaction and refine the deposit.

Blocking endogenous peroxidase. Peroxidase is present in peroxisomes and may be found in macrophages. A related enzyme, catalase, is present in red blood cells. Suppression of the activity of these enzymes is carried out at any stage before application of the peroxidase-linked reagent, usually with excess of the enzyme's substrate, hydrogen peroxide, in methanol (itself an inhibitor of peroxidase), buffer or simply water. Alternatives for particularly intransigent endogenous enzymes are the periodate-borohydride method (Heyderman, 1979) and for fragile cell cultures or cryostat sections, hydrogen peroxide in 70% methanol in phosphate-buffered saline, the milder sodium nitroferricyanide or phenyl hydrazine methods (Straus, 1971, 1972) or the nascent hydrogen peroxide method (Andrew and Jasani, 1987). Some of these methods are detailed in the Appendix (Section A.4).

Occasionally the hydrogen peroxide block can damage an antigen. In this case, the endogenous enzyme can be blocked at any stage of the reaction after application of the primary antibody and before application of the peroxidase-labelled reagent.

Diaminobenzidine (DAB) development *(Appendix, Section A.7.1)*. The method in most common use, which produces a dark brown, insoluble precipitate at the site of reaction, is that of Graham and Karnovsky (1966) using hydrogen peroxide as the substrate and diaminobenzidine (DAB) as the chromogen. The tetrahydrochloride of DAB is used, as it is more soluble than the free base. The reaction is complex, and a simplified version is shown in *Figure 4.1*.

This method is one of the most sensitive available and provides a permanent preparation with good contrast (*Plates 2 and 3*). There are several methods for intensifying the colour of the end-product (see Chapter 8) and it can be made electron-dense for electron microscopy by treatment with osmium tetroxide.

Safety note: There is evidence that prolonged exposure to benzidine, of which DAB is a derivative, can be carcinogenic in man (International Agency for Research on Cancer, 1972). It should be noted that evidence

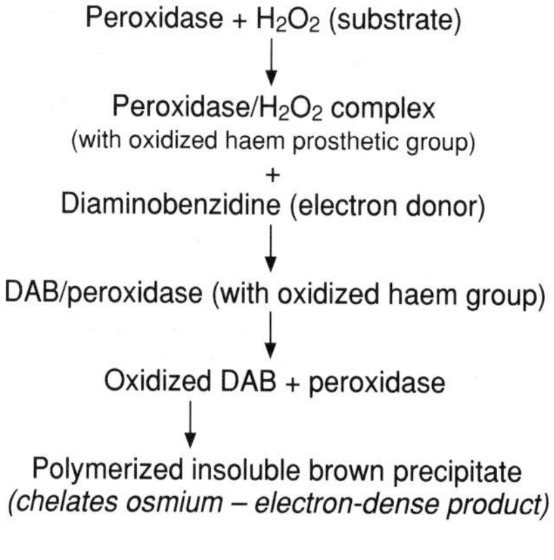

Peroxidase + H_2O_2 (substrate)
↓
Peroxidase/H_2O_2 complex
(with oxidized haem prosthetic group)
+
Diaminobenzidine (electron donor)
↓
DAB/peroxidase (with oxidized haem group)
↓
Oxidized DAB + peroxidase
↓
Polymerized insoluble brown precipitate
(chelates osmium – electron-dense product)

Figure 4.1: Simplified mechanism of peroxidase reaction with DAB and H_2O_2.

comes from subjects undergoing prolonged exposure to the compound during the industrial manufacturing process when concentrations are likely to be at far higher levels than would be encountered from the 50 mg% solutions used in histochemistry. Furthermore, the experiments of Weisburger *et al.* (1978) indicated that the addition to benzidine of the two amino groups to form DAB almost eliminated the carcinogenic effect relative to benzidine. Nevertheless, precautions to prevent undue exposure and contamination are advisable, such as keeping one area of the laboratory, preferably in a fume cupboard, for DAB development, wearing gloves when handling solutions of DAB, and treating the solution with sodium hypochlorite (ordinary household bleach is adequate) after it is finished with, as well as glassware, instruments and spills. Sodium hypochlorite oxidizes the DAB (the solution turns black) and reduces the toxicity. In the USA an alternative method of oxidation with excess acidified potassium permanganate is now the required method of disposal (Lunn and Sansome, 1990). (See Appendix, Section A.7.1.) After oxidation, the solution can be washed down the drain with plenty of water. DAB solutions are most safely stored frozen in aliquots (Pelliniemi *et al.*, 1980) (see Appendix, Section A.7.1).

3-Amino-9-ethylcarbazole *(Appendix, Section A.7.1).* Another chromogen, 3-amino-9-ethylcarbazole (AEC), was introduced as a less carcinogenic alternative to DAB (Graham *et al.*, 1965). Later work suggested that this substance too may be carcinogenic (Tubbs and Sheibani, 1982) and it is therefore wise to treat all chemicals as potentially toxic.

AEC with hydrogen peroxide as a substrate for the peroxidase reaction produces a bright red end-product *(Plate 4,* p. 34) which may be preferred to the brown of the DAB reaction. The product dissolves in organic solvents, so a water-based mountant must be used, and care must be taken not to differentiate a blue haematoxylin nuclear counterstain in acid-alcohol. Aqueous 0.1% hydrochloric acid could be used if necessary, or the problem avoided by using Mayer's haemalum as a progressive nuclear counterstain.

4-Chloro-1-naphthol *(Appendix, Section A.7.1).* A grey-blue reaction product, also alcohol-soluble, is produced with 4-chloro-1-naphthol as chromogen (Nakane, 1968), which may be useful in multiple immunostaining (see Chapter 9).

Phenol tetrazolium reaction *(Appendix, Section A.7.1).* The very sensitive phenol tetrazolium reaction (Murray *et al.*, 1991) produces a dark blue formazan precipitate. The pseudo-peroxidase of red blood cells does not take part in this reaction, but any endogenous true peroxidase will still require blocking.

The variety of colours that can be achieved has been enlarged further by the series of chromogens produced by Vector Laboratories. These can be very useful in immunostaining several different antigens in one preparation (see Chapter 9).

4.2.2   Alkaline phosphatase (Appendix, Section A.7.2)

This enzyme was originally introduced for quantitative work with the ELISA technique (see Chapter 9), providing a soluble, coloured end-product of reaction that could be measured with a spectrophotometer. However, the enzyme can also react in a technique using an azo dye as chromogen, producing a coloured precipitate, usually red or blue, suitable for immunocytochemistry (Burstone, 1961) (*Figure 4.2; Plate 5(a),* p. 35). The contrasting colours obtained with this marker are attractive and useful in double staining methods or where endogenous peroxidase or brown pigment is intrusive, but the end-products are often soluble and require aqueous mountants. Two development methods yield relatively solvent-fast products, either red (Malik and Damon, 1982) or blue-brown (de Jong *et al.*, 1985).

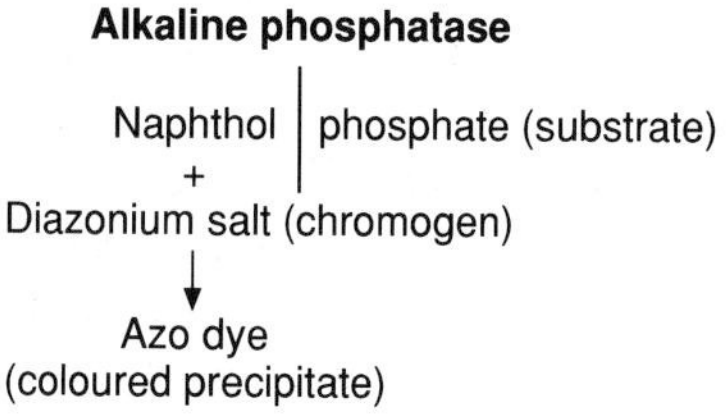

Alkaline phosphatase

Naphthol │ phosphate (substrate)
  +
Diazonium salt (chromogen)

↓

Azo dye
(coloured precipitate)

Figure 4.2: Simplified mechanism of action of alkaline phosphatase on a substrate and chromogen.

Blocking endogenous alkaline phosphatase. Alkaline phosphatase has several isoforms that are characteristic for different tissues such as bone, liver and endothelium. The isoenzyme used for immunolabelling is derived from (calf) intestine. All the isoforms except the intestinal one may be blocked by including 1 mM levamisole in the final enzyme development medium (Ponder and Wilkinson, 1981), but methods dependent on alkaline phosphatase as a label are not recommended for immunostaining intestinal tissue, particularly when fresh material is used. Most of the endogenous enzyme activity is destroyed during processing using paraffin, but some usually survives and can be confusing. It could be blocked with 1% acetic acid or 0.3 M hydrochloric acid, but this might damage some antigens.

4.2.3   Glucose oxidase (Appendix, Section A.7.3)

This is a plant enzyme, not present in animals. Glucose oxidase-labelled reactions are therefore very useful when endogenous peroxidase or alkaline phosphatase is a problem (e.g. fresh-frozen preparations of intestine), but are not suitable for immunocytochemistry on botanical preparations. The reaction product is dark blue and can be permanently mounted.

4.2.4   β-D-Galactosidase (Appendix, Section A.7.4).

This enzyme, derived from bacteria, differs from the mammalian form in its pH optimum. Therefore, provided that the correct pH is used for development, there is no need to consider reactions of the endogenous enzyme. The method would not be suitable for immunostaining preparations containing bacteria. The reaction usually used is the indigogenic one giving an insoluble, turquoise-blue product (*Plate 6*, p. 35).

4.3   Gold Labels

4.3.1   Colloidal gold

Gold labelling was first introduced for ultrastructural immunocytochemistry (Faulk and Taylor, 1971) because the electron-dense particles of colloidal gold are easily visible in the electron microscope; they remain the label of choice in this field (see Chapter 10), but are also useful for light microscopy. Gold particles are not conjugated to immunoglobulins in the same way as fluorochromes or enzymes, but rather, immunoglobulin molecules are adsorbed to the particles by non-covalent forces. Adsorption of protein is pH-dependent. It stabilizes the gold and is strongest at the iso-electric point of the protein. For a discussion and the preparation of gold-labelled reagents see the work by Roth (1983), De Mey (1986a,b) and Beesley (1989).

Colloidal gold is itself a pinkish-red colour and if enough reagent is applied (sometimes needing a build-up of several layers of gold-labelled antibody; see Chapter 8), the site of immunoreaction can be seen without further treatment. The particles reflect incident light, and if dark-field or epi-polarization microscopy is employed, the powerful back-scattering of reflected light means that highly diluted solutions of gold-labelled reagents can be used, which reduces both the expense of the reaction and non-specific binding to tissues.

Holgate *et al.* (1983a,b) greatly increased the sensitivity of the immunogold methods by adding an intensification step in which metallic silver was precipitated on to the gold particles by autometallography from silver lactate solution combined with a reducing agent, hydroquinone. Hacker *et al.* (1988) introduced silver acetate as a relatively light-insensitive alternative to silver lactate, which has made the method much more accessible (see Appendix, Section A.10).

Immunogold staining with silver intensification (IGSS) is one of the most sensitive immunocytochemical methods available. It has the advantage of producing an intensely black end-product which is very stable and can be combined with most other immunocytochemical or conventional staining procedures (*Plate 7*). None of the reagents is toxic and extremely high dilutions of primary antibodies can be used; this increases specificity and is economical. However, the method requires rigorous cleanliness and a considerable amount of expertise before it can be used routinely. The granular nature of the reaction product can be disturbing, and it is recommended that antibodies are adsorbed to the smallest possible size of gold particle (1–5 nm diameter) to minimize this effect and to allow easier penetration of labelled reagents into tissues (see Appendix, Section A.10).

4.3.2 Nanogold

Nanogold is even smaller than 1 nm diameter colloidal gold. The reagent consists of a core of gold atoms covalently attached to phosphorus atoms and surrounded by reactive groups (Hainfeld and Furuja, 1992). These allow it to be covalently coupled to proteins such as antibodies and strep-tavidin rather than being adsorbed to them like colloidal gold. This makes the complex more stable and because the gold is not exposed on the surface, there is less tendency for it to aggregate in solution. The small size of the nanogold gives labelled antibodies or other probes better penetrating power into tissues. It is therefore a very useful reagent for pre-embedding electron microscopical immunocytochemistry. Silver enhancement is required for both light- and electron microscopy, but the resolution is better than with colloidal gold. For a full discussion of gold and silver staining, see Hacker and Gu (2002).

Uses of immunogold staining. The IGSS method is useful when particularly high sensitivity is required, for instance when there is little antigen present. The intensity of the reaction product is so great that the immunolabelled structure can be seen easily at a low microscopic power. It has found a good use in haematology when immunolabelling is combined with Giemsa or another stain so that the labelled cells can be characterized easily, particularly when bright-field and epi-polarized illumination are combined (De Waele, 1989), as the silver deposit reflects light in the same way as the un-enhanced colloidal gold granules. The red colour of colloidal gold may be useful in multiple staining, especially as it does not require further histochemical development. The black, silver-intensified product provides yet another colour (see Section 9.3.3).

4.4   Other labels

4.4.1   Biotin

This is a small molecule, a vitamin, found in many tissues and extracted in quantity from the yolk of eggs. It is easy to attach a large number of biotin molecules to antibodies and, although it is not itself a visible label, it can be combined with any of the usual labels such as fluorochromes, enzymes or gold particles. However, usually the biotin attached to the antibody is not labelled but in a further reaction is allowed to combine with labelled avidin or an avidin-labelled biotin complex (ABC) (see Section 6.2.6). It could also be used as a hapten and located with an anti-biotin (see Section 4.4.2).

4.4.2   Haptens

An ingenious method for reducing non-specific binding of secondary antibodies to tissues is to label the primary antibody with a hapten. This is a substance that does not occur naturally in tissues, two examples being fluorescein and dinitrophenyl aminoproprionitrile imido ester (DNP). The primary antibody is detected by a second antibody raised to the hapten (Cammisuli and Wofsy, 1976; Jasani *et al.*, 1981). The second antibody may be labelled in any suitable way.

4.4.3   Radioisotopes

Because of the many problems inherent in the use of radioactive substances (danger, short half-life, the need for extended development times and poor resolution in autoradiographs), radiolabelled immunocytochemistry is rather specialized. It can be useful in semi-quantitative analysis of immunolabelling and in double-labelling with non-radioactive methods (for a review, see Hunt *et al.*, 1986). Isotopes can be incorporated into (monoclonal) antibody molecules (Cuello *et al.*, 1982), or an antibody may be made radioactive by reaction *in vitro* with an isotopically labelled antigen (Larsson and Schwartz, 1977). In this radioimmunocytochemistry (RICH) technique an isotope (^{125}I) is attached to an antigen which is then reacted with excess of its antibody. The specific antibody molecules are thus radiolabelled with the antigen, but still have one binding site free for attachment to another molecule of antigen in the tissue. The advantage of this method is that only specifically bound antibody will be detected in the subsequent autoradiograph. The disadvantages, in addition to the problems mentioned above, are that each antibody must be labelled individually and the technique of radiolabelling can be difficult. Enzyme-labelled antigen was used similarly by Mason and Sammons (1978), and gold-labelled antigen detection (GLAD) by Larsson (1979).

References

Andrew S, Jasani B. (1987) An improved method for the inhibition of endogeneous peroxidase non-deleterious to lymphocyte surface markers. Application to immunoperoxidase studies on eosinophil-rich tissue preparations. *Histochem. J.* **19**, 426–430.

Avrameas S. (1972) Enzyme markers: their linkage with proteins and use in immunohistochemistry. *Histochem. J.* **4**, 321–330.

Avrameas S, Uriel J. (1966) Méthode de marquage d'antigène et d'anticorps avec des enzymes et son application en immunodiffusion. *C. R. Acad. Sci. Paris Ser. D.* **262**, 2543–2545.

Beesley, JE. (1989). *Colloidal Gold: A New Perspective for Cytochemical Marking,* Microscopy Handbook 17. Oxford University Press, Oxford.

Bishop, AE, Polak JM, Bloom SR, Pearse AGE. (1978) A new universal technique for the immunocytochemical localisation of peptidergic innervation. *J. Endocrinol.* **77**, 25P–26P.

Bondi A, Chieregatti G, Eusebi V, Fulcheri E, Bussolati G. (1982) The use of β-galactosidase as a tracer in immunohistochemistry. *Histochemistry* **76**, 153–158.

Burstone MS. (1961) Histochemical demonstration of phosphatases in frozen sections with naphthol AS-phosphates. *J. Histochem. Cytochem.* **9**, 146–153.

Cammisuli S, Wofsy L. (1976) Hapten-sandwich labelling, III. Bifunctional reagents for immunospecific labelling of cell surface antigens. *J. Immunol.* **117**, 1695–1704.

Chu AC. (1986) Immunocytochemistry in dermatology. In *Immunocytochemistry, Modern Methods and Applications* (eds JM Polak, S Van Noorden). Butterworth–Heinemann, Oxford, pp. 618–637

Coons AH, Creech HJ, Jones RN. (1941) Immunological properties of an antibody containing a fluorescent group. *Proc. Soc. Exp. Biol. Med.* **47**, 200–202.

Coons AH, Leduc EH, Connolly JM. (1955) Studies on antibody production. I: A method for the histochemical demonstration of specific antibody and its application to a study of the hyperimmune rabbit. *J. Exp. Med.* **102**, 49–60.

Cowen T, Haven AJ, Burnstock B. (1985) Pontamine sky blue, a counterstain for background autofluorescence in fluorescence and immunofluorescence histochemistry. *Histochemistry* **82**, 205–208.

Cuello AC, Priestley JV, Milstein C. (1982) Immunocytochemistry with internally labeled monoclonal antibodies. *Proc. Natl Acad. Sci. USA* **79**, 665–669.

de Jong ASH, Van Kesse-Van Vark M, Raap AK. (1985) Sensitivity of various visualisation methods for peroxidase and alkaline phosphatase activity in immunoenzyme histochemistry. *Histochem. J.* **17**, 1119–1130.

De Mey J. (1986a) Gold probes in light microscopy. In *Immunocytochemistry, Modern Methods and Applications* (eds JM Polak, S Van Noorden). Butterworth–Heinemann, Oxford, pp. 71–88.

De Mey J. (1986b) The preparation and use of gold probes. In *Immunocytochemistry, Modern Methods and Applications* (eds JM Polak, S Van Noorden). Butterworth–Heinemann, Oxford, pp. 115–145.

De Waele M, Renmans W, Segers E, De Valck V, Jochmans K, Van-Camp B. (1989) An immunogold–silver staining method for detection of cell surface antigens in cell smears. *J. Histochem. Cytochem.* **37**, 1855–1862.

Evans DJ. (1986) Immunohistology in the diagnosis of renal disease. In *Immunocytochemistry, Modern Methods and Applications* (eds JM Polak, S Van Noorden). Butterworth–Heinemann, Oxford, pp. 638–649.

Falck B, Hillarp N-A, Thieme G, Torp A. (1962) Fluorescence of catecholamines and related compounds condensed with formaldehyde. *J. Histochem. Cytochem.* **10**, 348–354.

Faulk WR, Taylor GM. (1971) An immunocolloid method for the electron microscope. *Immunochemistry* **8**, 1081–1083.

Graham RC, Karnovsky MJ. (1966) The early stages of absorption of injected horse-radish peroxidase in the proximal tubules of mouse kidney: ultrastructural cytochemistry by a new technique. *J. Histochem. Cytochem.* **14**, 291–302.

Graham RC, Ludholm U, Karnovsky MJ. (1965) Cytochemical demonstration of peroxidase activity with 3-amino-9-ethylcarbazole. *J. Histochem. Cytochem.* **13**, 150–152.

Hacker GW, Gu J. (eds) (2002) *Gold and Silver Staining: Techniques in Molecular Morphology.* CRC Press, Boca Raton.

Hacker GW, Grimelius L, Danscher G, Bernatzky G, Muss W, Adam H, Thurner J. (1988) Silver acetate autometallography: an alternative enhancement technique for immunogold–silver staining (IGSS) and silver amplification of gold, silver, mercury and zinc in tissue. *J. Histotechnol.* **11**, 213–221.

Hainfeld JF, Furuja FR. (1992) A 1.4 nm gold cluster covalently attached to antibodies improves immunolabelling. *J. Histochem. Cytochem.* **40**, 177–184.

Heyderman E. (1979) Immunoperoxidase techniques in histopathology: applications, methods and controls. *J. Clin. Pathol.* **32**, 971–978.

Holgate CS, Jackson P, Cowen PN, Bird CC. (1983a) Immunogold–silver staining – new method of immunostaining with enhanced sensitivity. *J. Histochem. Cytochem.* **31**, 938–944.

Holgate CS, Jackson P, Lauder I, Cowen PN, Bird CC. (1983b) Surface membrane staining of immunoglobulins in paraffin sections of non-Hodgkin's lymphomas using an immunogold–silver technique. *J Clin. Pathol.* **36**, 742–746.

Hökfelt T, Ljungdhal Å. (1972) Modification of the Falk–Hillarp formaldehyde fluorescence method using the Vibratome: simple, rapid and sensitive localisation in sections of unfixed and formalin-fixed brain tissue. *Histochemie* **29**, 325–339.

Hunt SP, Allanson J, Mantyh PW. (1986) Radioimmunocytochemistry. In *Immunocytochemistry, Modern Methods and Applications* (eds JM Polak, S Van Noorden). Butterworth–Heinemann, Oxford, pp. 99–114.

International Agency for Research on Cancer (1972) *Some Inorganic Substances, Chlorinated Hydrocarbons, Aromatic Amines, N-Nitroso Compounds and Natural Products*, Monograph 1, Monographs on the Evaluation of the Carcinogenic Risk of Chemicals to Humans. IARC, Lyon, pp. 80–86.

Jasani B, Wynford Thomas D, Williams ED. (1981) Use of monoclonal anti-hapten antibodies for immunolocalisation of tissue antigens. *J. Clin. Pathol.* **34**, 1000–1002.

Johnstone A, Thorpe R. (1996) *Immunochemistry in Practice,* 3rd Edn. Blackwell Scientific Publications, Oxford.

Khalfan H, Abuknesha R, Rand-Weaver M, Price RG, Robinson D. (1986) Aminomethyl coumarin acetic acid: a new fluorescent labelling agent for proteins. *Histochem. J.* **18**, 497–499.

Larsson L-I. (1979) Simultaneous ultrastructural demonstration of multiple peptides in endocrine cells by a novel immunocytochemical method. *Nature* **282**, 743–745.

Larsson L-I., Schwartz TW. (1977) Radioimmunocytochemistry – a novel immunocytochemical principle. *J. Histochem. Cytochem.* **25**, 1140–1148.

Lunn G, Sansome E. (1990) *Destruction of Hazardous Chemicals in the Laboratory.* Wiley Interscience, New York.

Malik NJ, Damon ME. (1982) Improved double immunoenzymatic labelling using alkaline phosphatase and horseradish peroxidase. *J. Clin. Pathol.* **35**, 1092–1094.

Marrack J. (1934) Nature of antibodies. *Nature* **133**, 292–293.

Mason DY, Sammons RE. (1978) Alkaline phosphatase and peroxidase for double immunoenzymatic labelling of cellular constituents. *J. Clin. Pathol.* **31**, 454–462.

Murray GI, Foster CO, Ewen SWB. (1991) A novel tetrazolium method for peroxidase histochemistry and immunohistochemistry. *J. Histochem. Cytochem.* **39**, 541–544.

Nakane PK. (1968) Simultaneous localization of multiple tissue antigens using the per-oxidase-labeled antibody method: a study on pituitary gland of the rat. *J. Histochem. Cytochem.* **16**, 557–560.

Nakane PK, Pierce Jr GB. (1966) Enzyme-labeled antibodies: preparation and applica-tion for the localization of antigens. *J. Histochem. Cytochem.* **14**, 929–931.

Oi VT, Glazer AN, Stryer L. (1982) Fluorescent phycobiliprotein conjugates for analyses of cells and molecules. *J. Cell Biol.* **93**, 981-986.

Pelliniemi LJ, Dym M, Karnovsky MJ. (1980) Peroxidase histochemistry using diaminobenzidine tetrahydrochloride stored as a frozen solution. *J. Histochem. Cytochem.* **28**, 191–192.

Ponder BAJ, Wilkinson MM. (1981) Inhibition of endogenous tissue alkaline phos-phatase with the use of alkaline phosphatase conjugates in immunohistochemistry. *J. Histochem. Cytochem.* **29**, 981–984.

Riggs JL, Seiwald RJ, Burkhalter JH, Downs CM, Metcalf T. (1958) Isothiocyanate compounds as fluorescent labeling agents for immune serum. *Am. J. Pathol.* **34**, 1081–1097.

Roth J. (1983) The colloidal gold marker system for light and electron microscopic cyto-chemistry. In *Techniques in Immunocytochemistry,* Vol. 2 (eds GR Bullock, P Petrusz). Academic Press, New York, pp. 217–284.

Schenk EA, Churukian CJ. (1974) Immunofluorescence counterstains. *J. Histochem. Cytochem.* **22**, 962–966.

Scherbaum WA, Mirakian R, Pujol-Borrell R, Dean BM, Bottazzo GF. (1986a) Immunocytochemistry in the study and diagnosis of organ-specific auto-immune dis-eases. In *Immunocytochemistry, Modern Methods and Applications* (eds JM Polak, S Van Noorden). Butterworth–Heinemann, Oxford, pp. 456–476.

Scherbaum WA, Blaschek M, Berg PA, Doniach D, Bottazzo GF. (1986b) Spectrum and profiles of non-organ-specific auto-antibodies in auto-immune disease. In *Immunocytochemistry, Modern Methods and Applications* (eds JM Polak and S Van Noorden). Butterworth–Heinemann, Oxford, pp. 477–491.

Sternberger LA. (1986) *Immunocytochemistry*, 3rd Edn. John Wiley and Sons, New York.

Straus W. (1971) Inhibition of peroxidase by methanol and by methanol nitro-ferricyanide for use in immunoperoxidase procedures. *J. Histochem. Cytochem.* **19**, 682–688.

Straus W. (1972) Phenylhydrazine as inhibitor of horseradish peroxidase for use in immunoperoxidase procedures. *J. Histochem.Cytochem.* **20**, 949–951.

Suffin SC, Muck KB, Young JC, Lewin K, Porter DD. (1979) Improvement of the glu-cose oxidase immunoenzyme technique. *Am. J. Clin. Pathol.* **71**, 492–496.

Titus JA, Haughland R, Sharrows SO, Segal DM. (1982) Texas Red, a hydrophilic, red-emitting fluorophore for use with fluorescein in dual parameter flow microfluorimetric and fluorescence microscopic studies. *J. Immunol. Meth.* **50**, 193–204

Tubbs RR, Sheibani K. (1982) Chromogens for immunohistochemistry. *Arch. Pathol. Lab. Med.* **106**, 205.

Weisburger EK, Russfield AB, Homburger F, Weisburger JH, Boger E, Van Dongen CG, Chu KC. (1978) Testing of twenty-one environmental aromatic amines or derivatives for long-term toxicity or carcinogenicity. *J. Environ. Pathol. Toxicol.* **2**, 325–356.

Wessendorf MW, Appel NM, Molitor TW, Elde RP. (1990) A method for immunofluo-rescent demonstration of three coexisting neurotransmitters in rat brain and spinal cord, using the fluorophores fluorescein, lissamine rhodamine and 7-amino-4-methyl-coumarin-3-acetic acid. *J. Histochem. Cytochem.* **38**, 1859–1877.

5 Non-specific Staining due to Tissue Factors

Immunoreagents can unfortunately bind to tissue sites that do not contain their specific antigen. These non-specifically binding antibodies will be detected along with the specifically binding ones causing 'background' and confusing the final picture. This universal problem is discussed further in Chapter 7 in connection with antibody specificity, but it is worth understanding the causes that are due to common tissue factors at this stage because prevention of non-specific staining is an integral part of all immunocytochemical methods.

An important point is that all antibodies will bind unspecifically if they are highly concentrated, so all reagents should be diluted as far as possible, consistent with good immunostaining, to avoid unwanted binding as well as for economy. Polyclonal antibodies are usually more prone to causing background staining than monoclonals because they contain many types of potentially binding antibodies. However, monoclonals are not always blameless.

Negative control preparations should be included with every immunostaining run for every type of method. These are preparations in which the primary antibody has been left out or, preferably, replaced by an irrelevant antibody of the same species and immunoglobulin subclass. In all other respects the treatment is exactly the same as for the test preparation and there should be no stain seen at the end of the reaction. If any reaction is visible, it must be assumed that it is non-specific and attempts to prevent it can be made.

Fortunately, non-specific binding is weaker than true antigen–antibody binding, and there are some simple methods that can be used to overcome the common problems.

5.1   Causes of non-specific binding

5.1.1   Charged sites

Antibodies carry a positive or negative charge and can bind by electrostatic attraction to tissue sites with the opposite charge. Some tissue
components such as eosinophils carry a particularly strong positive
charge that can be hard to overcome, though it is reduced by aldehyde
fixation.

Raising the salt content of the diluent buffer to 2.5% increases its
ionic strength, which challenges the weak binding to charged sites.
Raising the pH to 9 has a similar effect as has adding a little detergent
to the rinsing buffer (0.05% Tween 20) (Buffa *et al.*, 1979; Grube, 1980).
These remedies may reduce the amount of specific binding of a polyclonal antibody by preventing binding of the weaker specific antibodies
of the immunoglobulin population; the standard dilution of the antibody
may have to be decreased to allow for this.

5.1.2   Hydrophobic attraction

Exclusion of water molecules through aldehyde fixation of tissues
encourages interactions between the hydrophobic sites and proteins
such as applied immunoglobulins. These interactions, at least for polyclonal antibodies, are discouraged by raising the pH of the diluent to 9
or adding detergent.

5.1.3   Fc receptors

In fresh tissue (frozen sections and cytological preparations) tissue
receptors for the Fc portion of antibodies may pose an additional
problem. These Fc receptors, present on several cell types such as
macrophages and monocytes as part of the natural immune defence
mechanism, are largely destroyed by formalin fixation and tissue processing. If necessary, F(ab) fragments of antibodies which lack the Fc
portion should be used.

5.2   Prevention of non-specific binding

There are two simple universal remedies for these tissue factor
problems. The first is to dilute the antibodies as far as possible and the
second is to block potential binding sites with concentrated protein
solution. Normal (non-immune) serum in high concentration (undiluted

to 1/20) is applied to the preparation before applying the specific primary antibody. Normal serum contains enough natural antibodies and other proteins to occupy all these binding sites and prevent attachment of the specific antibody. The species providing the normal serum is important, and depends on the method being used (see Chapter 6).

With the exception of Fc receptors in fresh tissue, non-specific binding sites can generally be blocked by an 'inert' protein such as bovine serum albumin, chicken egg albumen or casein (fat-free dried milk powder will do) in a 1 or 2% solution in buffer. These are much cheaper than serum. However, chicken egg albumin should be avoided if biotinylated reagents are to be used as it is a source of avidin and will attract the biotin (see Section 6.2.6.)

These non-specific bonds are much weaker than true antigen–antibody binding. They can generally be prevented by including detergent in the washing buffer (e.g. 0.05% Tween 20) or by raising the salt content of the antibody diluent from 0.9% to 2.5% (Buffa *et al.*, 1979).

Polyclonal antibodies should be used as highly diluted as possible to reduce the concentration of proteins that might bind non-specifically (*Plate 8*, p. 36). The danger is smaller with monoclonal antibodies as they contain only one immunoglobulin, unless they are derived from an ascites fluid which may contain native proteins from the host animal.

5.3   Other problems

5.3.1   Endogenous enzymes

Ways of inhibiting endogenous peroxidase and alkaline phosphatase have been discussed in Chapter 4.

5.3.2   Endogenous biotin

Tissue biotin can be blocked by unlabelled avidin and biotin as described in Chapter 6 (Section 6.2.6) and Appendix A.4.3 (Wood and Warnke, 1981). Biotin blocking reagents are available commercially, but an inexpensive 'home-made' version consists of skimmed or fat-free dried milk (milk is a rich source of biotin) and egg white (a rich source of avidin) (Miller and Kubier, 1997; Miller *et al.*, 1999). The method is given in Appendix A.4.3.

5.3.3   Autofluorescence

Some tissue components are intrinsically fluorescent, such as elastic tissue and lipofuschin. Usually the colour of the autofluorescence is different from that of the applied fluorochrome and is easily seen in a negative control preparation. If it is troublesome, a fluorescent counterstain may be useful (see Section 4.1.6) or a different fluorochrome.

References

Buffa R, Solcia E, Fiocca R, Crivelli, O, Pera A. (1979) Complement-mediated binding of immunoglobulins to some endocrine cells of the pancreas and gut. *J. Histochem. Cytochem.* **27**, 1279–1280.

Grube D. (1980) Immunoreactivities of gastrin (G) cells. II; Nonspecific binding of immunoglobulins to G-cells by ionic interactions. *Histochemistry* **66**, 149–167.

Miller RT, Kubier P. (1997) Blocking of endogenous avidin-binding activity in immuno-histochemistry: the use of egg whites. *Appl. Immunohistochem.* **5**, 63–66.

Miller RT, Kubier P, Reynolds B, Henry T, Turnbow H. (1999) Blocking of endogenous avidin-binding activity in immunohistochemistry; the use of skim milk as an economical and effective substitute for biotin. *Appl. Immunhistochem. Molec. Morphol.* **7**, 63–65.

Wood GS, Warnke R. (1981) Suppression of endogenous avidin binding activity in tissues and its relevance to biotin-avidin detection systems. *J. Histochem. Cytochem.* **29**, 1196–1204.

6 Methods

6.1 General considerations

Immunocytochemistry is a very tolerant technique. There are many satisfactory variants of the methods – those given here are practiced in the authors' laboratories. If results are assessed on the basis of whether a reaction is present or absent, timing need not be precise, over a minimum needed for reaction, as it should allow for saturation of the antigen at the optimal dilution of the antibody. If any kind of quantitative comparison is to be done, conditions must be uniform throughout the experiment.

6.1.1 Buffers (Appendix, Section A.1)

Immunocytochemical reactions traditionally take place in a buffer solution that stabilizes the antibody and does not damage the tissue substrate. The latter condition is more important for fresh than for fixed tissue. The pH and ionic content of buffers used for rinsing and, particularly, for diluting monoclonal antibodies may have an important influence on the efficiency of the immunoreaction. Binding to charged tissue sites is optimal when the isoelectric point of the environment is distant from that of the antibody and its antigen. Boenisch (1999) showed that the presence of sodium chloride in the diluent was actually deleterious in several instances and that Tris buffer gave better results than phosphate buffer. All the antibodies tested could be used satisfactorily in PBS or TBS at the dilutions suggested by the manufacturer, but greater efficiency (dilution) could be achieved by optimizing the pH for each individual antibody. Polyclonal antibodies, containing a wide range of immunoglobulins with different isoelectric points, were not affected by the parameters tested. In an experimental situation it would be worth testing a new monoclonal antibody at pH 6.0 and pH 8.0 in Tris buffer without sodium chloride, but for routine purposes it is difficult to arrange an individual pH diluent for each antibody, and efficiency may have to be sacrificed to expediency. In practice, the most commonly used buffers for diluting and for rinsing are 0.01 M phosphate-buffered 0.9% NaCl at pH 7.0–7.4 (phosphate-buffered saline, PBS) or 0.05 M Tris/HCl with 0.9% NaCl at pH 7.6 (Tris-buffered saline, TBS). The pH must be higher than 7.0 to prevent detachment of antibodies from the tissue, which becomes a danger at low pH, but it can be raised as high as 9.0 if necessary to prevent non-specific binding of reagents (Grube, 1980). It is said that the use of TBS rather than PBS carries a lower risk of non-specific background staining because of its high ionic strength, but in our experience there is little difference. PBS is cheaper than TBS, particularly if it is made in the laboratory. It is convenient to make a stock solution of about 10 l of 0.1 M phosphate buffer with 9% NaCl and dilute 1 l to 10 l for daily use. TBS is recommended for use with alkaline phosphatase-labelled methods and for diluting alkaline phosphatase-labelled reagents. In general, the buffer used for rinsing should also be the basis for the antibody diluent, but in practice it does not seem to matter greatly whether the same medium is used throughout or not. Detergent (e.g. 0.05% Tween 20 or for fresh material 0.1% saponin; see Section 3.4) may be added to the rinsing buffer to help prevent non-specific binding of antibodies (see Section 5.2), but this will reduce the surface tension of a glass slide so that a buffer rinse without detergent may be helpful at the stage before the antibody is applied.

6.1.2   Antibody diluent and storage

Storage of undiluted antibodies. Suppliers put an expiry date on their antibodies. This indicates the length of time for which they have been able to test the antibody and can guarantee its properties, but obviously they cannot test each antibody to extinction. Balaton and his colleagues (1999) did a great service to immunocytochemists in testing a wide variety of antibodies and showing that they could be used satisfactorily well beyond the expiry date, often for many years. There is certainly no need to throw away antibodies indiscriminately at the end of their given shelf-life. The important proviso is that when immunostaining quality of the positive control begins to decline, the antibody should be discarded. Undiluted antibodies should be stored according to the supplier's recommendations.

Storage of diluted antibodies. It is convenient and efficient to store some of each antibody at its working dilution to avoid the chore of making a new dilution every time the antibody is used and to provide greater uniformity between tests.

Many antibodies can be stored at their working dilution at 4°C for several weeks to years provided that a preservative such as sodium azide (0.01–0.1%) is included in the buffer diluent and also a protecting protein which will attach to any sticky sites on the walls of the storage vessel in competition with the highy diluted antibody. Bovine serum albumin at 0.1% is usually used. Neither of these additives interferes with immunocytochemical reactions. Buffer alone is adequate if the antibody is diluted just prior to use. It cannot be assumed that every antibody will withstand storage at its working dilution – this must be tested for each antibody. Alternative methods of storage are given in the Appendix (Section A.2).

Enzyme-linked reagents should not be diluted in buffers containing azide, since this substance inhibits many enzyme reactions. In general, it is wiser not to store any labelled antibodies at high dilutions as there is a slight danger that the label will become detached from the antibody. However, protective solutions for storing diluted peroxidase-linked reagents are now available (e.g. Protexidase from ICN), and for any laboratory in which immunocytochemistry is performed regularly it is certainly an advantage to have solutions pre-diluted in the refrigerator. Commercial kits providing ready-to-use diluted solutions will be adequately preserved.

6.1.3 Antibody dilution relative to reaction time, temperature and technique

Primary antibodies. The optimum dilution for an antibody is the highest at which specific immunoglobulin can saturate the available antigen, leaving some unbound antibody in the solution to ensure continued binding. It can only be established by testing the antibody in a series of dilutions on positive control tissue known to contain the antigen and prepared in the same way and labelled with the same method as the proposed experimental samples. Doubling-dilutions are convenient to use, usually starting with 1/50 for a polyclonal antiserum and going up to 1/6400 in the first instance (Section 7.1.3 and Appendix A.2.1). Further dilution may be possible, or the final dilution can be refined within the chosen range. Monoclonal antibodies are usually used at an antibody concentration of between 1 and 20 μg ml^{-1}. Provided the same method is used, the optimal dilution established on tissue known to contain a high quantity of antigen should be suitable for all antigen levels, but if the sensitivity of the method is increased, the dilution may need to be increased to avoid background staining.

Reaction time. Optimal dilutions are dependent on the time allowed for the reaction. A short reaction time (30 min) will require a lower dilution of the antiserum. This is because polyclonal antisera contain a mixture of antibodies which will differ in avidity for the antigen. In a short period, only the most avid of the antibody populations will bind to their targets. If reaction time is prolonged, dilution can be increased and all types of antibody molecule will have a chance to bind. Higher dilution is advantageous because the concentration of unwanted antibodies that may cause background staining will be reduced (*Plate 8*, p. 36).

Monoclonal antibodies by definition contain only one type of antibody. It may bind quickly or slowly, so if a short incubation period yields no reaction, it is worth extending it to four hours at room temperature or overnight at 4°C.

Secondary antibodies. The incubation time with commercial secondary antibodies, provided that they do not produce their own background staining, can generally be reduced to 30 min at room temperature. This is because they have been purified and are highly concentrated, relative to the amount of bound primary antibody. The optimal dilution for a labelled secondary antibody is usually in the range 1/20 to 1/100, but may be higher. It should be established by a dilution series against a known primary antibody used at its optimal dilution on a positive control. If the optimal dilution has been established for neither the primary nor the secondary antibodies, a chequer-board type of dilution experiment must be set up on known positive tissue with a range of dilutions for each antibody (*Table 6.1*) and the best combination chosen.

Temperature. The rate of immunological binding is also affected by temperature, being faster at 37°C than 27°C and slower at 4°C. It might be thought that a rapid reaction would be an advantage, but the rate of non-specific and unwanted binding will also be increased, so the danger of background staining with a high concentration of antiserum for a short period at 37°C is greater than with a high dilution overnight at 4°C. This consideration applies to whole antisera or total immunoglobulin fractions, but is less important with affinity-purified or monoclonal antibodies, which should be free of unwanted antibodies. Nevertheless, economy dictates the highest dilution possible. We have found that there

Table 6.1: Antibody titration using chequer-board dilutions (hypothetical) on positive control preparation to choose optimal dilutions of unknown primary and secondary antibodies in an indirect two-step method with standard incubation times

| Primary antibody | Secondary antibody | | | |
| | 1/100 | | 1/200 | |
	Cells	Background	Cells	Background
1/50	++++	++++	+++	+++
1/100	++++	++++	+++	+++
1/200	++++	+++	+++	++
1/400	+++	++	++	+
1/800	+++	+	++	±
1/1600	+++	−	+	−
1/3200	+	−	+	−
1/6400	−	−	−	−

In this hypothetical case the best dilution to choose would be 1/1600 for the primary antibody and 1/100 for the labelled secondary antibody.

is little difference in optimal dilution of a primary antiserum for an incubation period of 2 h at room temperature or overnight at 4°C – it is a matter of convenience which is used. The low temperature is used for prolonged incubation to prevent evaporation from the small drops of antiserum covering the preparation.

Technique. The optimal dilution of a primary antibody is also dependent on the sensitivity of the technique. For a two-step, indirect technique (see Section 6.2.4), the dilution will generally be about 10 times higher than for a one-step, direct technique in which it is the primary antibody that is labelled (see Section 6.2.3). The three-step methods allow a further five to 10 times higher dilution, because of their greater sensitivity (*Table 6.2*). *Table 6.2* refers to the classical standard immunoperoxidase techniques. However, there are highly labelled primary and secondary reagents that can raise the sensitivity of a one-step or two-step method impressively, allowing greater dilution. These reagents, EPOS™ and EnVision™ (Dako) and PowerVision™ (ImmunoVision Technologies) are described in Sections 6.2.3 and 6.2.4.

6.2 Methods

6.2.1 Nature of antibodies (IgG)

It is not necessary to be an immunologist to perform immunocytochemistry, but it is essential to know that most of the antibodies used in this technique are of the IgG class, though some may be IgM, and to understand the way in which the antibody molecules bind to their antigens. In the simplest terms, an IgG molecule consists of four polypeptide chains comprising two identical heavy chains making the constant fragment (Fc) of the IgG of a particular species, continuous with two combined

Table 6.2: Primary antibody dilution with hypothetical increase in sensitivity according to incubation time, temperature and method

Method	1 h at RT	Overnight at 4°C
Direct one-step	1/10	1/40
Indirect two-step	1/200	1/500
PAP three-step	1/500	1/2000
ABC three-step	1/1000	1/5000

heavy and light chains, which are variable and which each have one binding-site for the same particular antigen (antigen-binding fragment, F(ab)), selected and induced by immunization. The antigen-binding domain at the extreme end of the variable F(ab) fragment is known as Fv. The Fc portion carries the specific antigenic determinants to which antibodies raised to that particular IgG can bind. Because the polypeptide chain is in duplicate, there are at least two antigenic determinants on each Fc portion, as well as some on the F(ab) portions. IgG molecules are Y-shaped and are represented diagrammatically in this form (*Figure 6.1*). IgM molecules have a different configuration (for further details, see any immunology text book).

Immunocytochemical staining consists of applying a series of antibodies; the first one (primary) binds to the antigen in the tissue and then can itself act as an antigen for a second antibody, raised in an unrelated animal species to the IgG of the species providing the primary. The second antibody will bind to the antigenic sites on the Fc portion of the primary antibody molecule, and then may act as an antigen for a third antibody, and so on (see descriptions of methods below). The reagents in one of the layers will carry the label to identify the immunoreaction site.

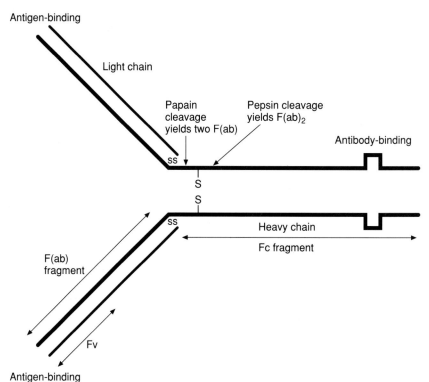

Figure 6.1: Simplified diagram of IgG molecule.

6.2.2   Application of antibodies to preparations

The equipment required for immunocytochemistry can be as simple as a Petri dish containing some wads of cotton wool or paper tissue dampened with distilled water to keep the atmosphere in the dish humid so that drops of solution do not evaporate from the preparation, and a rack made of wooden sticks or glass rods to support the microscope slide or coverslip carrying the section or cells. The dish should be covered during incubations. More elaborate incubating chambers with integral racks and washing channels are available commercially. Reagents can be applied with a Pasteur pipette or a pipette with disposable tips, and preparations can be rinsed with a wash-bottle or (preferably, because of the greater bulk of solution) in a slide carrier in a dish of buffer.

The traditional method of applying antibodies to sections on slides is to take the slide from buffer solution (e.g. PBS), dry it with a cloth or paper tissue leaving the section area damp, lay it in a humid chamber and drop a little antibody on to the section to cover it completely. Provided the section area is damp, the antibody will flow over the entire area, and if the slide has been carefully dried, the damp area will be isolated by the surface tension at the edge of the dried area. The absolute amount of antibody is not usually important, and depends on the area of the section. An average quantity is 100 μl per section. The slides must be kept horizontal to prevent the antibody droplet running off the section. It can be very helpful to encircle the section with a water-repellent pen. The pen must be applied to a dry slide; this is done most easily during hydration of a paraffin section at some stage after removing the paraffin with solvent and before it is taken to water. A cryostat section can be ringed after fixation in acetone, or before fixation in an aqueous fixative. Another advantage of the pen is that it allows separation of several sections on the same slide so that they can be stained with different antibodies or even, in an emergency, one section to be divided for different immunostains without danger of the solutions running together.

Another way of ensuring that antibody stays on the section is to cover the section and antibody droplet with a coverslip, but this is not usually necessary and adds considerably to the cost.

At the end of the incubating period, slides are drained briefly on to a tissue, making sure that if two antibodies are on the same slide the drops do not mingle on the sections. They are then placed in a slide carrier and immersed in buffer for rinsing. We have never seen spurious staining from the small quantities of different antibodies that must be present in the first rinse when this method is used.

After several changes of buffer, the slide is removed and dried around the section, ensuring that the surface of the retaining hydrophobic ring is dry. Excess buffer is drained off the section on to a tissue while keeping the section moist. If too much buffer is left on, the next antibody layer applied will become diluted beyond its optimum.

Automation. Several types of automated immunostainer are now available. They work either by incubating the slides horizontally and dropping or spraying antibodies and other reagents on to the preparation, or by holding the slides vertically against another slide or clip and allowing the solutions to be drawn up by capillarity to cover the preparation. In a further type, a drop of antibody spreads downwards and is held between the slide and the coverplate. A semi-automated version of this type uses the same system, but solutions must be dripped in by hand. This system is flexible, increases efficiency and uniformity of staining and is relatively inexpensive. It would benefit laboratories undertaking only small amounts of immunostaining

To be efficient, an automated method must be done in as short a time as possible. Antibody dilutions established for overnight incubations 'by hand' may have to be reduced to take into account a shorter incubation period on an immunostainer. We found that some primary antibodies needed a higher than normal concentration of second and third reagent; this complicates the programming. Some companies supply sets of primary and secondary reagents optimised for their instruments. As with all standard kits, this is inflexible and expensive, but provides the important advantage of simplicity.

In all the fully automated versions, application of antibodies depends on a mechanical arm programmed by computer to apply the correct volume of the correct solution at the correct time, with suitable rinses between applications. Much time and effort in washing and wiping slides can be saved by these machines in a busy laboratory, but they are still only as good as their programmers, who must know what they are doing in terms of immunocytochemistry and be aware of problems.

6.2.3   Direct (one-step) method

In this, the simplest of the immunocytochemical methods, the reaction is a one-step process with a labelled primary antibody. In the original method (Coons and Kaplan, 1950) the label conjugated to the antibody was fluorescein isocyanate. The conjugated antiserum, diluted in PBS, was allowed to react with a tissue section and the unbound antibody was then washed off with PBS. The section was examined in an ultraviolet microscope and the site of attachment of the antibody fluoresced apple-green. Many other labels have since been used, including other fluorochromes, enzymes, colloidal gold and biotin. Biotin would require a further reaction with labelled avidin.

A series of antibodies useful in diagnostic immunocytochemistry, the EPOS™ (Enhanced Polymer One Step) is produced by Dako. These allow much greater sensitivity than a conventional peroxidase-labelled primary antibody because they are conjugated to a dextran polymer chain that is very highly labelled with the enzyme, providing an intense reaction on the antigenic site in the tissue.

Blocking. Before application of the primary antibody, possible non-specific tissue binding sites are blocked by application of normal serum from the same species as the primary antibody (although any species will do, provided that the antigen to be localized is not immunoglobulin which might lead to cross-reactions; see Chapters 5 and 7). The blocking serum is usually diluted 1/20 in PBS or TBS diluent (see Section 6.1.2), but a higher concentration may be used if necessary. This will occupy Fc receptors and hydrophobic and electrostatic binding sites. The reactions are weak, so the serum is not washed off the tissue but merely drained off, after a minimum time of about 10 min, and the labelled antibody is then applied, optimally diluted in diluent (see *Figure 6.2*).

It is important to use serum for blocking background in fresh-frozen tissue or cell preparations because it will contain enough immunoglobulin to occupy the Fc receptors as well as other non-specific binding sites. In paraffin sections the Fc receptors are inactivated and the serum could be replaced by a protein such as bovine serum albumin or casein that will block non-Fc receptor sites effectively.

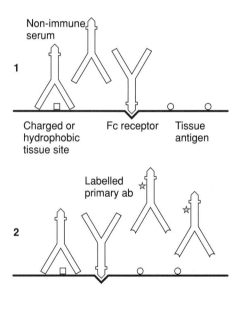

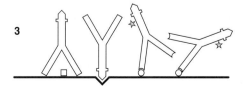

Figure 6.2: Direct method.

6.2.4 Indirect (two-step) method

Coons and his colleagues increased the sensitivity of their immunofluorescence method by introducing a second layer (Coons *et al.*, 1955). The first layer is the primary antibody raised to the antigen to be localized. This is not itself labelled, but is detected after binding by a labelled secondary antibody raised to the immunoglobulin of the species that donated the first antibody. Thus, if the primary antibody is a rabbit IgG anti-antigen, the secondary antibody might be a goat (or swine or sheep) anti-rabbit IgG, labelled with FITC or any other label (*Figure 6.3*).

Blocking. In all methods employing two or more layers of immunoreagent, the background-blocking normal serum (immunoglobulin)

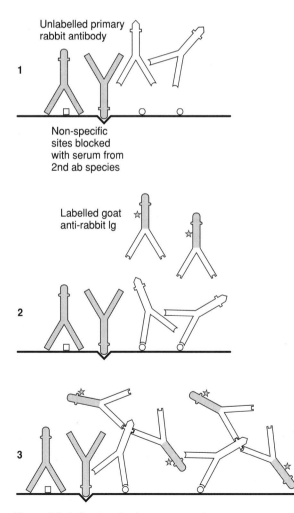

Figure 6.3: Indirect method.

applied before the primary antibody must be from the species that provides the second antibody. If serum from the same species as the first antibody is used, it will effectively block non-specific binding of the first antibody, but will provide extra deposits of immunoglobulin antigen for the second, anti-species immunoglobulin, thus creating background staining. Serum from the second species will block non-specific binding of the primary antiserum equally well and cannot be immunoreactive to the immunoglobulin of its own species.

Advantages

1. Anti-IgG sera used as the second layer are usually made hyperimmune and of very high avidity.
2. At least two labelled anti-immunoglobulin molecules can bind to each primary antibody molecule, increasing the sensitivity of the reaction or detectability of the antigen.
3. Economy; one labelled second-layer antibody can be used to detect any number of first layer antibodies to different antigens, provided they have all been raised in the species donating the IgG against which the second layer antibody is directed.

Note that the primary antibody may be an IgM, particularly if it is a monoclonal antibody. In this case, the second layer antibody must be either an anti-IgM or a mixture of antibodies to the entire immunoglobulin fraction of the species providing the primary antibody. These are usually described in catalogues as anti-immunoglobulins. Anti-IgG alone will not react with an IgM primary. It might be necessary to decrease the working anti-IgG dilution of a secondary antibody to all immunoglobulins to detect an IgM primary, as the concentration of antibodies to IgM may be less than to IgG.

Sensitivity up to the level of a three-step labelled avidin method can be achieved in a two-step reaction with Dako's EnVision™ reagents. These are secondary antibodies conjugated to polymers that carry a large number of peroxidase or alkaline phosphatase molecules (Vyberg and Nielsen, 1998). A similar reagent, with a more compact polymer chain is the PowerVision™ reagent from ImmunoVision Technologies (Daly City, California, USA) (Shi *et al.*, 1999).

6.2.5 Three-layer methods

Peroxidase–anti-peroxidase (PAP) or unlabelled antibody–enzyme method. A further development of the indirect technique led to the exceedingly sensitive double-immunoglobulin bridge (Mason *et al.*, 1969) and the unlabelled antibody–enzyme or peroxidase–anti-peroxidase (PAP) methods (Sternberger *et al.*, 1970).

This technique involves yet a third layer, which (for a rabbit primary antibody) is a rabbit antibody to peroxidase, coupled with peroxidase in such proportions that it forms a cyclic and very stable peroxidase–anti-peroxidase complex composed of two rabbit IgG molecules combined with three peroxidase molecules, one of which they share. The PAP complex acts as a third-layer antigen and becomes bound to the unconjugated (goat) anti-rabbit IgG of the second layer. This must be in excess with respect to the first layer, so that its antibody molecules compete for the antigen-bound primary rabbit IgG. The competition means that only one of the two antigen-binding sites of each second-layer antibody molecule is occupied by the primary antibody, now acting as an immunoglobulin antigen, and the second site is free to combine with the PAP complex, another rabbit immunoglobulin antigen (*Figure 6.4*).

Advantages. This method results in 100–1000 times higher sensitivity than the indirect method for several reasons. The peroxidase molecule is not chemically conjugated to the antibody but immunologically bound, so it loses none of its enzyme activity. Similarly, the antibodies are all unencumbered by conjugated label and retain their full activity. In addition, much more peroxidase ends up on the site of reaction than with the indirect method. The increased sensitivity arising from this allows for much higher dilution of the first antibody, thus eliminating many of the unwanted components of the primary antiserum and reducing non-specific staining. The reaction is particularly specific since the PAP complex is highly purified and will react only with the anti-rabbit

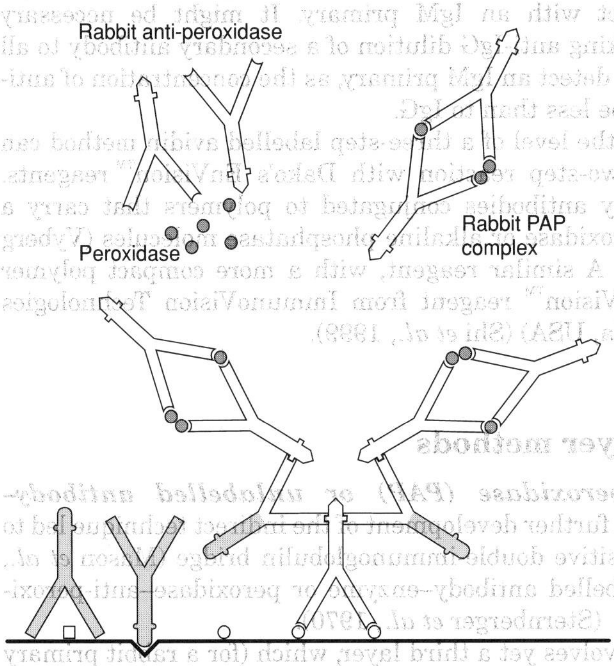

Figure 6.4: Peroxidase–anti-peroxidase method.

immunoglobulin of the second layer. It will not react with the blocked tissue, nor with non-specifically attached immunoglobulins from the blocking layer since these are not anti-rabbit IgG, so background staining is minimal. PAP complexes are made with IgG from many species, for example mouse and goat, as well as rabbit.

Disadvantages. The only disadvantage of the three-layer techniques is that they require an extra reagent and increased time, and may be more expensive than the two-layer method, depending on the dilutions used. The advantages of greater sensitivity and higher dilution of the primary antiserum outweigh the disadvantages.

Alternative enzyme–anti-enzyme complexes. Alkaline phosphatase–anti-alkaline phosphatase (APAAP) can be made more simply than PAP complexes by mixing alkaline phosphatase in excess with monoclonal mouse anti-alkaline phosphatase (Cordell *et al.*, 1984). APAAP complexes are not cyclic like PAP but merely consist of the maximal amount of alkaline phosphatase antigen bound to each molecule of alkaline phosphatase, so the ratio of antigen to antibody is a little less

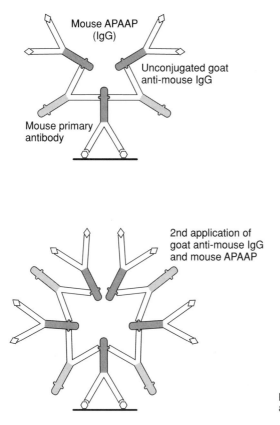

Figure 6.5: Alkaline phosphatase–anti-alkaline phosphatase method.

than for PAP. To increase the sensitivity of the APAAP method, Mason (1985) recommended a second application of the anti-mouse IgG after the APAAP layer followed by a second application of APAAP. Because APAAP is an IgG molecule, the anti-IgG will bind to it and the second application of APAAP will bind to spare F(ab) of the anti-IgG. This provides more label on the reaction site (*Figure 6.5*). Mouse APAAP is an IgG_1 complex so it can be used only if the primary antibody is also a mouse IgG_1. An IgM primary antibody cannot be used with this method.

APAAP complexes are also available made in rat and rabbit. Complexes are also available made with glucose oxidase.

6.2.6   Avidin–biotin methods

Even greater sensitivity in terms of the amount of label finally bound to the antigenic site is produced by the avidin–biotin methods. Avidin is a large glycoprotein extracted from egg white (albumen). It has four binding sites per molecule for a low molecular weight vitamin called biotin and a particularly high affinity for it (hence the avoidance of egg albumen as a blocking medium (see Section 5.4). Biotin is a vitamin found in several tissues and extracted (conveniently) in bulk from egg yolk. Each biotin molecule has one binding site for avidin and can be attached through other sites to an antibody or any other macromolecule such as an enzyme, fluorophore or other label. Avidin, too, may be labelled, so these reagents can be used in a variety of immunostaining techniques. The increased sensitivity results from the large number of biotin molecules that can be attached to an antibody and the large amount of label or biotinylated label that can be combined with avidin (*Plate 9*, p. 37) (Guesdon *et al.*, 1979; Hsu and Raine, 1981; Hsu *et al.*, 1981; Coggi *et al.*, 1986).

Labelled avidin method. In this method the second antibody is biotinylated and the third reagent is avidin labelled with peroxidase or other marker. Because of the large number of biotin molecules attached to the antibody, many labelled avidin molecules may be bound at the site of the primary antigen–antibody reaction (*Figure 6.6*).

Avidin–biotin complex (ABC) method. In this variant, the second antibody is biotinylated and the third reagent is a complex of avidin mixed with biotin that has been labelled with an appropriate marker. The avidin and labelled biotin are allowed to react together for at least 30 min before being applied, resulting in the formation of a large and highly labelled complex, the label molecules being shared by several biotin molecules. The proportion of avidin to labelled biotin must be such that some binding sites on the avidin are left free to attach to the biotin on the second antibody (see *Figure 6.7*). Manufacturers supply solutions

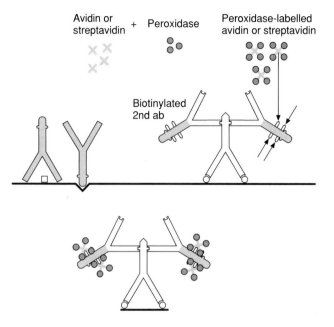

Figure 6.6: Labelled (strept)avidin method.

that are adjusted to give the correct ratio. This method is said to be more sensitive than the labelled avidin method because of the increased amount of label in the larger complex.

Disadvantages. There is a danger that the complex can become too big and the reaction suffer from steric hindrance as the ABC is prevented by its size from reaching the biotin on the antibody. The manufacturer usually minimizes this effect by distancing the biotin from the antibody with a spacer arm and by judicious balancing of the avidin and biotinylated label. Avidin may produce background staining by binding to lectins in the tissue through its carbohydrate groups and also through charged reactions since its isoelectric point is 10. These problems are overcome either by using an avidin specially modified by the supplier or by substituting it with streptavidin. Streptavidin is very similar to avidin except that it is a protein, not a glycoprotein, so avoiding the lectin-binding possibility, and it has a neutral isoelectric point, thus avoiding charged reactions. It is extracted from the bacterium *Streptomyces avidinii.*

Some tissues such as kidney and liver are rich in biotin, which can produce unwanted reactions with labelled avidin or the ABC, particularly after heat-mediated antigen retrieval, which may reveal hidden biotin (Bussolati *et al.*, 1997). Endogenous biotin can be blocked easily by applying unconjugated avidin. This is then saturated with biotin. Because each biotin molecule has only one avidin-binding site, this procedure effectively blocks unwanted attachment of the ABC or labelled avidin (Wood and Warnke, 1981). Solutions may be made from

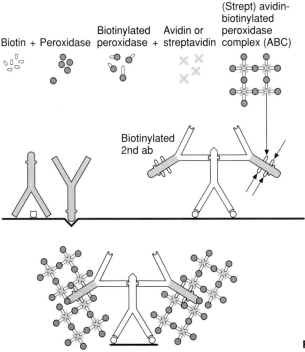

Figure 6.7: ABC method.

avidin and biotin, or kits of these solutions are available commercially. As mentioned in Section 5.3.2, blocking solutions made from egg white (avidin) and skimmed or powdered milk (biotin) are inexpensive and usually effective (Miller and Kubier, 1997; Miller *et al.*, 1999). (*Figure 6.8* and *Plate 10* and Appendix, Section A.4.3).

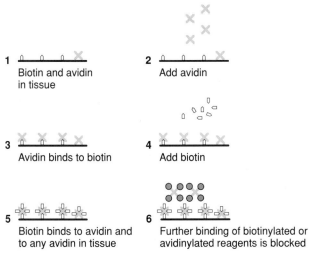

Figure 6.8: Blocking endogenous biotin.

References

Balaton AJ, Drachenberg CB, Rucker C, Vaury P, Papadimitriou JC. (1999) Satisfactory performance of primary antibodies beyond manufacturers' recommended expiration dates. *Appl. Immunhistochem. Molec. Morphol.* **7**, 221–225.

Boenisch T. (1999) Diluent buffer ions and pH: their influence on the performance of monoclonal antibodies in immunohistochemistry. *Appl. Immunohistochem. Molec. Morphol.* **7**, 300–306.

Bussolati G, Giuliotta P, Volante M, Pace M, Papotti M. (1997) Retrieved endogenous biotin: a novel marker and a potential pitfall in diagnostic immunohistochemistry. *Histopathology* **31**, 400–407.

Coggi G, Dell'Orto P, Viale G. (1986) Avidin–biotin methods. In *Immunocytochemistry, Modern Methods and Applications* (eds JM Polak, S Van Noorden). Butterworth–Heinemann, Oxford, pp. 54–70.

Coons AH, Kaplan MH. (1950) Localization of antigen in tissue cells. *J. Exp. Med.* **91**, 1–13.

Coons AH, Leduc EH, Connolly JM. (1955) Studies on antibody production. I: A method for the histochemical demonstration of specific antibody and its application to a study of the hyperimmune rabbit. *J. Exp. Med.* **102**, 49–60.

Cordell JL, Falini B, Erber WN, Ghosh AK, Abdulaziz Z, MacDonald S, Pulford KAF, Stein H, Mason DY. (1984) Immunoenzymatic labeling of monoclonal antibodies using immune complexes of alkaline phosphatase and monoclonal anti-alkaline phosphatase (APAAP complexes). *J. Histochem. Cytochem.* **32**, 219–222.

Grube D. (1980) Immunoreactivities of gastrin (G) cells. II; Nonspecific binding of immunoglobulins to G-cells by ionic interactions. *Histochemistry* **66**, 149–167.

Guesdon JL, Ternynck T, Avrameas S. (1979) The uses of avidin–biotin interaction in immunoenzymatic techniques. *J. Histochem. Cytochem.* **27**, 1131–1139.

Hsu SM, Raine L. (1981) Protein A, avidin and biotin in immunocytochemistry. *J. Histochem. Cytochem.* **29**, 1349–1353.

Hsu SM, Raine L, Fanger H. (1981) Use of avidin–biotin–peroxidase complex (ABC) in immunoperoxidase techniques: a comparison between ABC and unlabeled antibody (PAP) procedures. *J. Histochem. Cytochem.* **29**, 577–580.

Mason DY. (1985) Imunocytochemical labeling of monoclonal antibodies by the APAAP immunoalkaline phosphatase technique. In *Techniques in Immunocytochemistry,* Volume 3 (eds GR Bullock, P Petrusz). Academic Press, New York, pp. 25–42.

Mason TC, Phifer RF, Spicer SS, Swallow RA, Dreskin RB. (1969) An immunoglobulin–enzyme bridge method for localizing tissue antigens. *J. Histochem. Cytochem.* **17**, 563–569.

Miller RT, Kubier P. (1997) Blocking of endogenous avidin-binding activity in immunohistochemistry: the use of egg whites. *Appl. Immunohistochem.* **5**, 63–66.

Miller RT, Kubier P, Reynolds B, Henry T, Turnbow H. (1999) Blocking of endogenous avidin-binding activity in immunohistochemistry; the use of skim milk as an economical and effective substitute for biotin. *Appl. Immunhistochem. Molec. Morphol.* **7**, 63-65.

Shi S-R, Guo J, Cote RJ, Young L, Hawes D, Shi Y, Thu S, Taylor CR. (1999) Sensitivity and detection efficiency of a novel two-step detection system (PowerVision) for immunohistochemistry. *Appl. Immunohistochem. Mol. Morphol.* **7**: 201–208.

Sternberger LA, Hardy Jr PH, Cuculis IJ, Mayer HG. (1970) The unlabeled antibody–enzyme method of immunohistochemistry. Preparation and properties of soluble antigen–antibody complex (horseradish peroxidase–anti-horseradish peroxidase) and its use in identification of spirochetes. *J. Histochem. Cytochem.* **18**, 315–333.

Wood GS, Warnke R. (1981) Suppression of endogenous avidin-binding activity in tissues and its relevance to biotin–avidin detection systems. *J. Histochem. Cytochem.* **29**, 1196–1204.

Vyberg M, Nielsen S. (1998) Dextran polymer conjugate two-step visualization system for immunohistochemistry: a comparison of EnVision+ with two three-step avidin–biotin techniques. *Appl. Immunohistochem.* **6**, 3–10.

7 Testing Antibodies: Specificity and Essential Controls

Every antibody must be tested to establish the optimum conditions for its use, which will vary according to the immunostaining method and may, therefore, differ from those suggested in the literature or data sheet. However, recommended procedures should be used as a guide. A positive-appearing result may well be genuine, but there is a risk of non-specific or unwanted specific reactions which must be eliminated before the result can be accepted. These reactions are due to tissue factors, which have been discussed in Chapter 5, and to antibody factors, and should be tested for in parallel with the specific reaction, including all pre-treatment and blocking steps.

Some common problems and their remedies are listed in *Table 7.1*. Some less common causes of non-specific or unwanted staining are discussed in work by Van Noorden (1993).

7.1 Testing a new primary antibody

Use secondary detecting reagents of known optimal dilution for your method.

7.1.1 A primary antibody with a known localization

1. Read the data sheet and published literature about the antibody and start by following the conditions recommended. However, remember that your own laboratory conditions and staining schedule may be different and this could affect the result.
2. Obtain tissue known to contain the antigen and prepared in the same way as your test samples. Ensure that you have enough preparations for all the dilutions and negative control.

3. Use a series of doubling dilutions of the primary antibody to immunostain the preparation by the chosen method, following any recommendations for pretreatment, such as protease digestion or antigen retrieval by heat. The procedure for making doubling dilutions has sometimes puzzled newcomers to the concept, so it is described in Appendix A.2.1.

    Polyclonal antisera. If a dilution is suggested, try a range about that dilution including one above and one below. If no dilution is suggested, begin with 1/50 as the lowest and use doubling dilutions (1/100, 1/200, 1/400 etc.) up to 1/6400 (8 dilutions).

    Purified polyclonal immunoglobulin. If no dilution is suggested, start at 40 µg ml^{-1}. Doubling dilutions will give 20, 10, 5, 2.5, 1.25, 0.675 µg ml^{-1} (7 dilutions). The concentration of antibody for immunocytochemistry is usually between 1 and 10 µg ml^{-1}.

    Monoclonal antibodies. If no dilution is suggested, start at 40 µg ml^{-1}. Doubling dilutions will give 20, 10, 5, 2.5, 1.25, 0.675 µg ml^{-1} (7 dilutions).

    Negative control. Include one negative control of the same preparation, replacing a primary polyclonal antibody with normal (non-immune) serum from the same species as the primary antibody at the usual negative control level (see Section 7.1.3) or an inappropriate polyclonal antibody of the same species whose antigen is absent from the preparation. Replace a monoclonal antibody with antibody diluent alone or with an inappropriate antibody of the same species and immunoglobulin subclass. The negative control will show the level of non-specific background staining that might be expected from that particular tissue with the method used and in the absence of the test antibody (see Section 7.2).

4. Assess the results. A dilution series should, ideally, start with a very strong reaction, perhaps with some background staining due to

---

Table 7.1: Problems and remedies

High level of background staining

1. Dilute the primary antibody further (assuming standard dilution for other reagents)

2. If this is ineffective, check negative control with second (and third) reagents only. If background is absent, the staining must be due to a reaction between the primary and the tissue which is detected by the secondary antibody. If background is still present, it could be due to:
(a) Tissue factors (non-specific binding sites, Fc receptors, basic proteins; see Chapter 5)
Remedy: (i)   Increase concentration of blocking protein/serum
           (ii)   Add detergent to buffer rinse (e.g. 0.05% Tween 20)
           (iii) Raise sodium chloride content of antibody diluent to 2.5%
           (iv)   Raise pH of buffer to 9.0
           (v)   (For Fc receptor binding) use F(ab) portions of antibodies
           (vi)   Add 2 mg poly-L-lysine (a basic protein) (mol. wt 3000–6000) to each ml of diluted antibody
(b) Endogenous biotin (in an ABC method)
Remedy: (i)   Block endogenous biotin (see Section 6.2.6 and Appendix A.4.3)
           (ii)   Use a non-ABC method

Table 7.1: Problems and remedies – *continued*

(c) Avidin binding to mast cells and other charged sites
Remedy: (i) Raise buffer pH to 9.0
(ii) Use streptavidin or 'modified' avidin instead of avidin
(d) Incompletely blocked endogenous peroxidase
Remedy: (i) Try longer/stronger blocking with H_2O_2
(ii) Use another enzyme label
(e) Cross-reaction between anti-species immunoglobulin (second antibody) and host-tissue immunoglobulins
Remedy: (i) Absorb cross-reacting antibody with 1% of host-tissue species normal serum or immunoglobulin
(ii) Use species-specific antibodies
(f) Aldehyde groups in tissue left from fixative
Remedy: (i) Wash tissue well before processing and embedding
(ii) Treat preparation with freshly made 0.02–1% sodium or potassium borohydride in 0.1 M phosphate buffer or water for 2–30 min at room temperature
(iii) Add 10–100 mM ammonium chloride to the blocking serum
(g) Antigen diffused due to inadequate fixation
No remedy
(h) Staining in necrotic areas due to damaged cell membranes. Necrotic cells can take up antigens from the environment
No remedy. Be aware of the state of the tissue

False-positive staining

(a) Positive-appearing stain due to endogenous pigment. Check negative control. The pigment will be there also
Remedy: (i) After developing a peroxidase-DAB immunoreaction, formalin pigment can be bleached with alcoholic picric acid, and mercury pigment with Lugol's iodine and sodium thiosulphate. Melanin can be bleached before doing the immunostain.
(ii) Use a red (AEC) reaction for peroxidase instead of DAB, or an alkaline-phosphatase label with red development. Alternatively, after DAB development, stain the melanin dark blue-green with azure B (Kligora *et al.*, 1999) or with Giemsa for 2 minutes followed by haematoxylin (Miller, 2001).
(b) Positive staining due to uptake of antigen by non-producing cells, e.g. macrophages
No remedy. Be aware that this can happen

Immunostaining weak or absent

1. Method sensitivity is insufficient for small quantity of antigen present
Remedy: (i) Increase sensitivity (see Chapter 8)

2. Antigen is hidden (over-fixed)
Remedy: (i) Antigen retrieval by protease treatment
(ii) Heat-mediated antigen retrieval

3. Antigen under-fixed (*Figure 7.1*). The antigen may diffuse and dissolve
No remedy. Be aware of the state of the tissue and avoid sampling error

4. Antibody deterioration
Remedy: (i) Check all primary antibodies against known positive controls with known positive second and third reagents
(ii) Check second antibodies against primaries known to be working well

5. Wrong antibody sequence applied in error
Remedy: (i) Rescue if possible by reapplication of antibodies (see Section 8.1.2)
(ii) Start again

6. Error in preparation of developing solutions (or any other reagent)
Remedy: (i) Start again

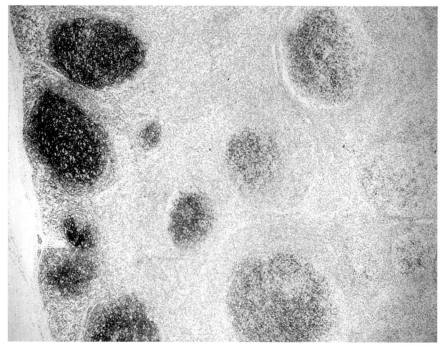

Figure 7.1: Human tonsil, formalin-fixed, paraffin embedded. This shows the effect of under-fixation. Pretreatment: microwave heating for 20 minutes in 0.01M citrate buffer, pH 6.0. Immunoperoxidase stain for CD 23, a marker of follicular dendritic cells. Note the decline in intensity of staining of the germinal centres from the well-fixed edge across the section, probably due to insufficient preservation of the antigen in the centre of the block.

concentrated antibody. As dilution increases, the background should diminish to zero and the specific staining should become more distinguishable. Eventually, the specific staining should lessen as the antibody concentration in the staining solution fails to saturate the antigen in the preparation. The dilution to choose would be one or two before the specific staining starts to diminish – i.e. one which still contains enough antibody to saturate abundant antigen. This dilution can be used as the standard dilution for this type of preparation unless the method is changed, when reassessment will be necessary.

5. If staining is still strong at the highest dilution, repeat with an extended series.

6. If the background is high with the test antibody, but not with the negative control, provided that there is some specific staining, try to find and cure the cause – see *Table 7.1*.

7. If there is no specific staining or it is very weak, try a longer period of antigen retrieval or a different method. Test buffers of low and high pH for heat-mediated antigen retrieval, or different proteases. Try a more sensitive method (see Chapter 8). If paraffin sections are failing

to react, try frozen sections as it may be that the antibody does not react on formalin-fixed tissue. In this case, try modifying the fixative.

8. Once the conditions are established, test the antibody on different tissues. For antibodies to be used in histopathological diagnosis, it can be useful and efficient to test antibodies on sections of multi-tissue blocks containing a variety of normal and tumour tissues, both positive and supposedly negative for the antigen concerned. Some companies supply sections of multiple tissue arrays.

7.1.2 A primary antibody with an unknown localization

Proceed in the same way as above, choosing preparations thought to contain the antigen. Specific-looking staining should be checked for cross-reactivity (see Section 7.2) and the specificity tested with absorption controls (see Section 7.2.4 and Appendix A.13).

7.1.3 Negative control for polyclonal antibodies – normal serum

When the primary antibody is contained in a polyclonal antiserum, a test for non-specific binding to the tissue being examined is to use in parallel, as the first layer, normal serum from a non-immunized animal of the same species that provides the primary antibody. The dilution of the normal serum, particularly if it is the pre-immune serum from the individual animal that provides the antibody, should theoretically be the same as for the primary antiserum being tested. However, at low dilutions (1/50–1/300) many non-immune sera contain enough immunoglobulins to provide significant background binding under immunocytochemical conditions, so the non-immune serum is used at the lowest dilution that has previously been shown to give no staining on a variety of preparations. This might be thought of as cheating but, even if the concentrations in the antiserum of immunoglobulins specific and non-specific to the antigen in question were known, which would be unusual, the selection of a suitable dilution for the primary antiserum is also experimental and subjective, chosen to give supposedly specific staining and no background (see Section 7.1.1). Thus any non-specific binding of immunoglobulins due to factors in the tissue being tested should occur with both the experimental antiserum and the normal serum at their chosen dilutions and be picked up by the anti-immunoglobulin of the second layer. Staining produced only with the antibody may well be specific.

7.1.4 Negative controls for monoclonal antibodies

Monoclonal primary antibodies are less prone to non-specific binding since, by definition, they contain only one antibody, but an irrelevant monoclonal antibody of the same immunoglobulin sub-class could be used instead of the primary to check the binding properties of the tissue sample. Failing that, substitution of culture medium (if the monoclonal is being used in this form) or antibody diluent could be used.

Further controls for both polyclonal and monoclonal antibodies would be to use the antibodies on tissue known not to contain the antigen, prepared in the same way as the test tissue. No staining should occur.

7.1.5 Testing for non-specific binding of second and third reagents

The presence of unwanted binding of second- and third-layer reagents to the tissue is easily determined by substituting antibody diluent alone for the primary antibody. In a three-layer reaction, the test should be carried out with both the second- and third-layer reagents and with the third-layer reagent alone.

7.2 Non-specific or unwanted specific staining due to antibody factors

7.2.1 Unwanted specific staining of unknown antigens

A polyclonal antiserum contains a mixture of antibodies, including the host animal's circulating antibody population. Some of these antibodies may react with unknown antigens in the tissue. These reactions, though specific, are likely to be less prominent than those of the antibodies resulting from immunization and are minimized by high dilution of the primary antiserum.

7.2.2 Non-specific binding of antisera to basic proteins

Scopsi *et al.* (1986) showed that some antisera and reagents such as streptavidin, particularly when labelled, bind to basic amino acids in tissue proteins, giving the impression of specific staining. Part of establishing whether a reaction is specific or not is to add a basic amino acid such as *poly*-L-lysine (molecular weight 3000–6000, 2 mg ml^{-1}; Sigma P0879) to the antibody at its working dilution to absorb any such non-specific reactivity before demonstrating the specific immunoreaction.

7.2.3 Unwanted specific cross-reactivity of anti-immunoglobulins

Because of the similarities of immunoglobulin molecules from different species, it is possible that the second-layer anti-immunoglobulins will react with the immunoglobulins of the animal in which the antigen is to be immunostained, for example anti-rabbit immunoglobulin may react with human immunoglobulin or (more likely) anti-mouse immunoglobulin with rat immunoglobulin. The immunoreactivity can be detected on sections of spleen from the antigen-containing animal. These may be fresh-frozen, acetone-fixed or formalin-fixed with heat-mediated antigen retrieval. Positive staining of plasma cells (which produce immunoglobulins) indicates that the antiserum contains cross-reacting anti-immunoglobulins. This staining can be eliminated by adding 1% of normal serum from the animal species to be tested to the working dilution of the cross-reacting antiserum. The effectiveness of the treatment can be tested again on a section of spleen. The working dilution of the antiserum may need to be reduced after addition of the normal serum because some of the antibody populations will have been neutralized, reducing the level of immunoreactive antibodies remaining. Species-specific anti-immunoglobulins are available commercially in which cross-reacting antibodies have already been absorbed with immunoglobulins from a number of species.

The potential for immunoglobulin cross-reactivity has also been exploited commercially in the provision of second-layer antisera which will react with primary antibodies raised in both rabbit and mouse (and sometimes other species as well). This is convenient as it avoids the need for deciding which second layer is appropriate to a stain – but it precludes double immunostaining with antibodies raised in the two species (see Chapter 9).

Mouse antibodies on mouse tissues. It is sometimes necessary to immunostain mouse tissues with a mouse monoclonal antibody. Unless the primary antibody is directly labelled or biotinylated, this presents the intractable problem that the second antibody (anti-mouse immunoglobulin) may reveal native mouse immunoglobulins as well as the primary mouse antibody. Several companies now provide kits to overcome this difficulty (e.g., Vector Laboratories' 'Mouse-on-Mouse' blocking kit and Dako's Animal Research Kit (ARK) for combining the primary antibody with biotinylated F(ab) fragments of secondary antibody).

7.2.4 Cross-reactivity of the primary antibody with related antigens

The same amino acid sequences occur in molecules of peptides belonging to the same family, and an antiserum raised by immunization with one member of the family may well contain antibodies reactive with the common sequences, resulting in failure to discriminate between the related peptides. For example, immunocytochemical staining of the human large intestine with antibodies to pancreatic polypeptide (PP) showed cells apparently containing the peptide, though none could be found in extracts of that area. It was not until the discovery of another peptide (with tyrosine at the N- and C-terminals, PYY) that it was realized that the PP antiserum was cross-reacting with PYY, with which it shared amino acid sequences (Ali-Rachedi *et al.*, 1984).

If only some epitopes are involved, it may be possible to remove the cross-reactivity by absorption of the antiserum with the related antigen. However, a monoclonal antibody may be just as cross-reactive as a polyclonal one if it happens to react with a shared epitope, and this reactivity cannot be absorbed out without removing all the immuno-reactivity. The possibility of specific cross-reactivity with known or unknown related antigens should be borne in mind, particularly when experimenting with a new antibody.

Testing for cross-reactivity with related antigens. If the related antigens are available, an enzyme-linked immunosorbent assay (ELISA) test against any number of antigens is probably the simplest way of establishing cross-reactivity. Dot blotting and Western blotting are other *in vitro* methods which will allow the reactions of an antibody with pure antigens or antigens in a tissue extract to be investigated. These methods are described briefly in Chapter 11.

Absorption with antigen (Appendix, Section A.13). An essential control with new antibodies and new tissues is addition of excess of the pure antigen to the antiserum. This should result in lack of staining, and if the reaction is specific, addition of related or unrelated antigens should have no effect.

In order to ensure excess of antigen over antibody, the sub-optimal dilution of the antibody should be used, giving less than maximal but consistent staining of all positive structures on control tissue. If possible, the antigen quantities added to the diluted antibody solution should be measured in nanomoles per millilitre of antibody, so that different antigens and antibodies can be compared. It is important to use pure antigen rather than the immunogen, which may have been coupled with a carrier protein. If the antibodies to the carrier protein were in fact responsible for the perceived immunoreactivity, absorption with the immunogen might remove the staining, giving a false impression that it was specific for the antigen itself. A suitable amount of antigen to add is 10–20 nmol ml^{-1} of sub-optimally diluted antibody; this should be adequate to remove immunoreactivity by binding to all the available antibody molecules. A stepwise 10-fold decrease in the ratio of antigen to antibody down to 0.001 nmol ml^{-1} should result in the gradual resumption of staining.

Positive control tissue should be tested alongside the experimental tissue to ensure that the absorption is effective.

7.3 Remedies for non-specificity due to tissue factors

7.3.1 Blocking binding sites with normal serum

This has been described in Chapter 5 (Section 5.2) and should remove binding to charged or hydrophobic tissue sites and to Fc receptors.

7.3.2 Absorption with tissue powder

Non-specific attachment of immunoglobulins to common tissue components can also be prevented by absorption of the antiserum with a tissue powder (e.g. acetone-dried liver) from the species in which the staining is to be done, provided it is certain that the tissue powder does not contain the antigen to be investigated. After reaction with the tissue powder, the mixture is centrifuged and the supernatant used as the antibody. It may be necessary to re-establish the optimal dilution.

7.4 Remedies for non-specificity due to heterogeneity of the antibody

7.4.1 Dilution

High dilution of the antiserum can reduce the amount of unwanted antibody compared with the amount of wanted, specific antibody in the serum to the point where its effects become negligible.

7.4.2 Affinity purification

Antisera can be 'purified' by immunoabsorption, with the specific antigen bound to a solid phase such as sepharose beads: the antibody is subsequently eluted. However, the most useful antibodies for immunocytochemistry are very avid and it may be difficult to elute them from the antigen used for absorption. Thus some antibody may be lost and the eluted antibody, though pure, may be of low avidity.

If the contaminating antibody is known, it may be possible to remove it from solution by solid phase absorption with the contaminant, the specific antibody passing unaffected through the column.

7.5 Remedies for non-specificity due to cross-reactivity

Genuine cross-reactivity is a difficult problem. In a polyclonal anti-serum, when relationships between families of antigen molecules are known, it may be possible to remove populations of cross-reacting antibodies by absorption with the related antigens, leaving some non-cross-reacting antibodies still available in the serum.

Another approach is to produce a non-cross-reacting, region-specific antibody. This requires immunization with unshared fragments of the respective molecules.

7.6 Essential staining controls

7.6.1 Negative controls

The conditions for these have already been discussed, but it must be stressed that a negative control must be performed for every tissue block or sample stained and for every method used in every run. For example, if the same tissue preparation is being immunostained with an indirect, two-layer immunoperoxidase method for one antigen and an APAAP method for another, then there must be included a negative control for each method. These are identical to the test preparations except for the substitution of normal serum, inappropriate antibody or buffer for the primary antibody. Even if it has already been shown that the preparation under test is negative with the method used, it is advisable to include one to allow for unforeseen variations in the reagents or method. If any staining occurs on the negative control, preferably the cause should be investigated and overcome, but at the very least, the result must be mentally subtracted from the result of the test.

7.6.2 Positive controls

It is essential that a positive control sample, known to contain the antigen in question, is included for every antibody in a run using the same method as for the test preparations. This must be done every time an immunostain is performed. Without such a control, a negative result on the test material will be meaningless, because there is no guarantee that the reagents are in good working condition and have been applied in the correct order and at the correct dilutions. If the positive control is satisfactory, it is a reasonable assumption that the correct method was carried out on the test material too. If the control is weaker than usual or unstained, then it is likely that something has gone wrong and the test result is unreliable.

7.6.3   Experimental controls

When an unknown tissue is being tested with an unknown antibody to show an antigen in an unknown location, seemingly positive results must be accepted with extreme caution. Absorption controls with the antigen are essential, and preferably with related and unrelated antigens and basic amino acids (see Sections 7.2.2 and 7.2.4) as well. If several different antibodies to the same antigen localize it in the same place, this constitutes confirmatory evidence of its presence.

References

Ali-Rachedi A, Varndell IM, Adrian TE, Gapp DA, Van Noorden S, Bloom SR, Polak JM. (1984) Peptide YY (PYY) immunoreactivity is co-stored with glugagon-related immunoreactants in endocrine cells of the gut and pancreas. *Histochemistry* **80**, 487–491.

Kligora CJ, Fair KP, Clem MS, Patterson JW. (1999) A comparison of melanin bleaching and azure blue counterstaining in the immunohistochemical diagnosis of malignant melanoma. *Mod. Pathol.* **12**, 1143–1147.

Miller R. (2001) Technical Immunohistochemistry: achieving reliability and reproducibility of immunostains. www.appliedimmuno.org. Website of the Society for Applied Immunohistochemistry.

Scopsi L, Wang BL, Larsson L-I. (1986) Unspecific immunocytochemical reactions with certain neurohormonal peptides and basic peptide sequences. *J. Histochem. Cytochem.* **34**, 1469–1475.

Van Noorden S. (1993) Problems and solutions. In *Immunocytochemistry, A Practical Approach* (ed JE Beesley). Oxford University Press, Oxford, pp. 208–239.

8 Increasing Sensitivity and Enhancing Standard Methods

The sensitivity of an immunoreaction and the enhancement of a reaction product are different but related subjects. The sensitivity of a method may have to be increased to maximise the visibility of a very small amount of antigen, and this is the true sense of the term. Scopsi and Larsson (1986) measured the sensitivity of various types of immunoperoxidase stain on series of dot blots of descending quantities of antigen, using a uniform dilution of primary antibody. In practice, this degree of accuracy is not required for comparing different methods. Method sensitivity may be measured roughly by its efficiency – the highest dilution of the primary antibody achieved on a preparation containing abundant antigen before the stain begins to decline.

The impact of immunocytochemical stains comes from the contrast between the colour of the end-product and the unstained or counterstained background. Enhancing the visibility of the end product of a reaction will not increase its sensitivity in terms of detecting more antigen. The purpose may be to emphasise the presence of a small amount of antigen, to provide good conditions for photography or image analysis, or merely for aesthetic reasons. There are numerous methods for enhancing the intensity of the result, the aim being to achieve more marker at the site of reaction or to darken an existing product. All require the initial reaction to have little or no background staining, since this, too, will be intensified. The positive and negative controls should also be enhanced if comparison is needed.

8.1 Increasing sensitivity

8.1.1 Immunogold with silver enhancement

The immunogold reaction with silver intensification is one of the most sensitive methods. It has been described in Chapter 4 (Section 4.3 and Appendix A 10).

8.1.2 Build-up methods

The greater sensitivity of three-layer methods over two-layer methods in achieving more label at the site of the original antigen–antibody reaction has already been shown in Chapter 6. Any immunoglobulin can act as a bridging antigen for another antibody, and thus the layers of antibody and labelled antibody can be continued. Adding to a rabbit PAP reaction with a further layer of anti-rabbit Ig and, another rabbit PAP quadruples the amount of peroxidase available (*Figure 8.1*). If the anti-rabbit Ig is peroxidase-conjugated, even more peroxidase molecules are added to the site.

Build-up by doubling the application of the second and third layers has been recommended as routine for APAAP staining (Cordell *et al.*, 1984; Mason, 1985). This has been described in Chapter 6 (Section 6.2.5).

In a two-layer indirect method, after the labelled secondary antibody has been added, normal serum containing immunoglobulin from the species providing the primary antibody or another layer of the primary antibody can be applied to give further binding sites for another layer of the labelled second antibody (*Figure 8.2*). The mirror image complementary antibody (MICA) system (Mangham and Isaacson, 1999) is based on this method. Multiple layers of peroxidase-conjugated and unconjugated

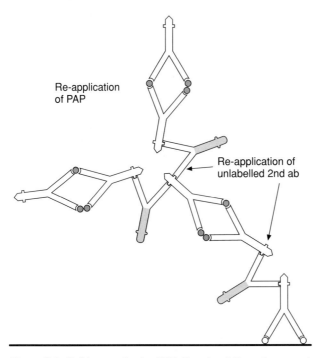

Re-application
of PAP

Re-application of
unlabelled 2nd ab

Figure 8.1: Build-up method – PAP. For simplicity, only one set of re-applied antibodies is shown. Note that a similar reaction will be present on the other side of the primary antibody molecule, resulting in a possible 24 peroxidase molecules after the build-up, instead of the original six.

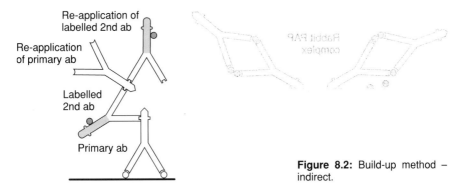

Figure 8.2: Build-up method – indirect.

antibodies are applied, resulting in great sensitivity exceeding that of the ABC method and avoiding problems due to endogenous biotin. The disadvantage is the extra time and the increased number of layers required. A kit called PolyMica™ is marketed by The Binding Site, Birmingham.

ABC methods are a little more difficult to intensify by increasing the numbers of layers because the reactive sites on the avidin and biotin are usually saturated. It may be necessary to change systems and follow the ABC with an anti-biotin and a labelled detecting antibody (*Figure 8.3*). Alternatively, if excess of the biotinylated second antibody is used at the first application, a PAP can be applied either with or after the peroxidase-labelled avidin or ABC to increase the amount of peroxidase at the antigenic site (*Figure 8.4*).

Saving failed reactions. Note that build-up methods can be used to 'save' a failed reaction in many instances. For example, if the peroxidase development of an initial rabbit PAP reaction had failed, additional layers of anti-rabbit Ig and PAP would provide a second chance, assuming that the primary antibody had been correctly applied. Repetition of the sequence in a two-layer technique woud have a similar effect, and in both cases if the failure had been due to the wrong species second layer, such as anti-mouse Ig instead of anti-rabbit Ig, reapplication of the

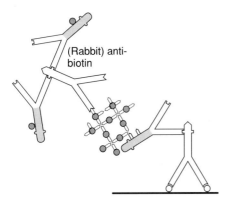

Figure 8.3: Enhancement of ABC method with anti-biotin.

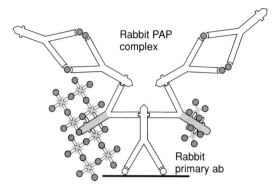

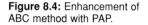

Figure 8.4: Enhancement of ABC method with PAP.

second (and third) layer(s) would restore the reaction, although not intensify it (*Figure 8.5*).

Addition of PAP might save a failed ABC reaction, provided that the biotinylated second antibody was originally in excess of the primary, so that spare immunoglobulin-binding sites were available. Alternatively, anti-biotin and another detection system might be used.

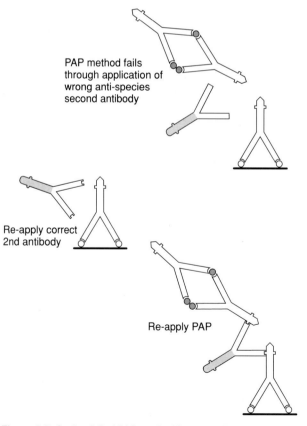

Figure 8.5: Saving failed PAP method by re-application of reagents.

8.1.3 Tyramine signal amplification (TSA)

This is also known as catalysed reporter deposition (CARD) and catalysed signal amplification (CSA). The method may be applied to any chromogen or fluorochrome in an ABC or labelled avidin reaction, provided that a peroxidase label is present on the original reaction. The amplifying power of the reaction is such that primary antibodies can be diluted greatly and usually have to be to avoid background staining, and sometimes pre-treatment with enzymes or heat can be avoided where it is normally essential to reveal an antigen. This method has made it possible to use some antibodies on paraffin sections that were previously only useful on frozen sections (*Figure 8.6*).

The usual immunostaining procedure is followed up to the end of the biotinylated second antibody stage. An extra layer of blocking protein is then applied to ensure adequate protein on the preparation. The next layer is the usual ABC or avidin (or streptavidin) labelled with peroxidase. After rinsing, the preparation is treated with biotinylated tyramine in the presence of hydrogen peroxide. The tyramine is activated by the catalytic action of the applied peroxidase and becomes bound to proteins in the tissue near the site of the peroxidase. The activation is thought to be due to the production of free oxygen radicals by the action of peroxidase on H_2O_2. The short duration of the activation (about 10 sec; P. De Jong, 1996, personal communication) means that the extra biotin is deposited only around the antigenic site. The final layer, after rinsing, is another application of ABC or labelled avidin/streptavidin, which binds to the extra biotin attached to the tyramine (*Figure 8.7*).

The initial peroxidase reagent could be a peroxidase-conjugated anti-immunoglobulin or PAP or even a peroxidase-labelled primary antibody rather than the biotinylated anti-immunoglobulin and avidin- or ABC-peroxidase, and the final layer could be a labelled anti-biotin, although these methods would be less sensitive that those containing avidin–biotin reactions. Provided that a peroxidase-labelled reagent has been applied, there is no need for the label of the final layer to be peroxidase – it can be any enzyme, fluorochrome or gold, and will still result in a 10- to 1000-times greater sensitivity than the standard ABC reaction. Kits for performing the amplification are available commercially (Du Pont NEM, Dako), but it is quite simple to biotinylate tyramine in the laboratory (Adams, 1992; Merz *et al.*, 1995; King, 1997).

Increased background staining may be a problem. Adams (1992) points out that it may be necesary to dilute the second antibody as well as the primary, or to use a species-specific one in order to prevent background staining from previously undetected binding of the second layer. It is also necessary to monitor the time in the standard DAB/H_2O_2 to avoid over-development; 1–3 min may be enough. Alternatively, adjust the concentrations of the antibodies and other reagents for the standard development time. As with most immunocytochemical methods, the

(a)

(b)

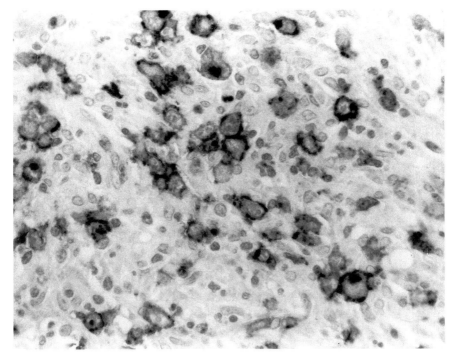

(c)

Figure 8.6: Lymph node from a case of Hodgkin's lymphoma. Formalin-fixed, paraffin-embedded sections immunostained with a peroxidase-labelled streptavidin system using a monoclonal primary antibody to CD 30, a marker of the characteristic Reed-Sternberg cells found in this disease. Haematoxylin counterstain. (a) Pretreatment: microwave heating in 0.01M citrate buffer, pH 6.0. Primary antibody at a dilution of 1/20. The Reed-Sternberg cells can just be seen (arrows). (b) No antigen retrieval. The primary antibody was used at 1/100 with tyramine signal amplification. The Reed-Sternberg cells are obvious. (c) Antigen retrieval as in (a). Primary antibody at 1/2000 dilution, with tyramine signal amplification. The combination of the two intensification methods allows a 1000-fold increase in dilution of the primary antibody.

end-point is subjective and the concentrations of all the reagents can be modified to provide a good result. The colour modification of the peroxidase/DAB/H_2O_2 end-product with cobalt or nickel salts (Section 8.2), or development with AEC or the phenol tetrazolium method (see Section 4.2), may also be applied to the TSA peroxidase product.

TSA is a very powerful amplification technique and there is a possibility that it may reveal hitherto unsuspected locations of antigens or minor degrees of cross-reactivity which were not recognized with the normal method. The usual controls should be applied and any unexpected results investigated.

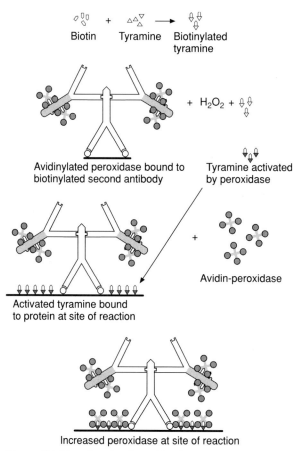

Figure 8.7: Tyramine signal amplification.

8.2 Intensification of the peroxidase/ DAB/H₂O₂ product

8.2.1 Post-reaction intensification (Appendix, Section A.8.1)

To intensify a brown DAB stain that is lighter than desired, the reaction product may be darkened with a variety of heavy metal salts. The simplest and least expensive is copper sulphate, which results in a dark brown colour with little effect on the unstained part of the preparation (*Plate 11*, p. 38; Hanker *et al.*, 1979). An alternative is a weak solution of gold chloride, which gives a blacker colour to the DAB product but tends to colour the rest of the preparation a reddish-brown. The reaction must be monitored under the microscope and may be adequate in a few seconds. It gives the stain a rather granular appearance. There is a commercially available DAB intensifying solution (Vector Laboratories) which is convenient to use and seems to have a similar effect. Several methods of intensifying with silver salts have been proposed (Gallyas *et al.*, 1982; Peacock *et al.*, 1991). They work efficiently but are more complicated than the copper or gold methods and may stain argyrophilic components of the tissue, which could be confusing (Gallyas and Wolff, 1986).

8.2.2 Intensification during the peroxidase reaction (Appendix A.8.2)

If a darker reaction product than that provided by DAB alone is required, it is more efficient to achieve it during the initial enzyme development than afterwards and less likely to produce enhanced background staining. Addition of salts of heavy metals such as cobalt or nickel to the incubation medium produce a dark blue-black or black product (see *Plates 4* and *12(c)*) (Hsu and Soban, 1982; Shu *et al.*, 1988). Imidazole is an alternative additive to the DAB incubating solution (Straus, 1982), producing a darker brown product but less intense than can be achieved with cobalt or nickel. Methods of intensifying the immunoperoxidase reaction were reviewed by Scopsi and Larsson (1986).

References

Adams JC. (1992) Biotin amplification of biotin and horseradish peroxidase signals in histochemical stains. *J. Histochem. Cytochem.* **40**, 1457–1463.

Cordell JL, Falini B, Erber WN, Ghosh AK, Abdulaziz Z, MacDonald S, Pulford KAF, Stein H, Mason DY. (1984) Immunoenzymatic labeling of monoclonal antibodies using immune complexes of alkaline phosphatase and monoclonal anti-alkaline phosphatase (APAAP comlexes). *J. Histochem. Cytochem.* **32**, 219–222.

Gallyas F, Görcs T, Merchenthaler I. (1982) High-grade intensification of the end-product of the diaminobenzidine reaction for peroxidase histochemistry. *J. Histochem. Cytochem.* **30**, 183–184.

Gallyas F, Wolff JR. (1986) Metal-catalyzed oxidation renders silver intensification selective. *J. Histochem. Cytochem.* **12**, 1667–1672.

Hanker JS, Ambrose WW, James CJ et al. (1979) Facilitated light microscopic cytochemical diagnosis of acute myelogenous leukemia. *Cancer Res.* **39**, 1635–1639.

Hsu SM, Soban E. (1982) Color modification of diaminobenzidine (DAB) precipitation by metallic ions and its application to double immunohistochemistry. *J. Histochem. Cytochem.* **30**, 1079–1082.

King G, Payne S, Walker F, Murray GI. (1997) A highly sensitive detection method for immunohistochemistry using biotinylated tyramine. *J. Pathol.* **183**, 237–241.

Mangham DC, Isaacson PG. (1999) A novel immunohistochemical detection system using mirror image complementary antibodies (MICA). *Histopathology* **35**, 129–133.

Mason DY. (1985) Immunocytochemical labeling of monoclonal antibodies by the APAAP immunoalkaline phosphatase technique. In *Techniques in Immunocytochemistry*, Vol. 3 (eds GR Bullock, P. Petrusz). Academic Press, New York, pp. 25–42.

Merz H, Malisius R, Mannweiler S, Zhou R, Hartmann W, Orscheschek K, Moubayed P, Feller AC. (1995) Methods in laboratory investigation. Immunomax. A maximized immunohistochemical method for the retrieval and enhancement of hidden antigens. *Lab. Invest.* **73**, 149–156.

Peacock CS, Thompson IW, Van Noorden S. (1991) Silver enhancement of polymerised diaminobenzidine: increased sensitivity for immunoperoxidase staining. *J. Clin. Pathol.* **44**, 756–758.

Scopsi L, Larsson L-I. (1986) Increased sensitivity in peroxidase immunocytochemistry. A comparative study of a number of peroxidase visualization methods employing a model system. *Histochemistry* **84**, 221–230.

Shu S, Ju G, Fan L. (1988) The glucose oxidase–DAB–nickel method in peroxidase histochemistry of the nervous system. *Neurosci. Lett.* **85**, 169–171.

Straus W. (1982) Imidazole increases the sensitivity of the cytochemical reaction for peroxidase with diaminobenzidine at a neutral pH. *J. Histochem. Cytochem.* **30**, 491–493.

9 Multiple Immunostaining

It is often useful to be able to demonstrate more than one antigen in the same preparation with differently coloured or fluorescent end-products. When the primary antibodies are produced in different species, the method presents no difficulty. The main problem occurs when they are produced in the same species, in which case cross-reaction of second antibodies with the 'wrong' primary can occur. Another general consideration is whether the antigens are located in different tissue structures, requiring contrasting colours, or in the same structure, in which case conditions for colour mixing must apply. Useful discussion and many helpful tips can be found in van der Loos (1999).

9.1 Double direct immunostaining with separately labelled primary antibodies

In this method, the antibodies can be from the same or different species. Double (or multiple) immunofluorescence staining can be carried out very simply on a single preparation with no possibility of confusing cross-reactions if two or more primary antibodies are labelled with different fluorochromes, which can be viewed with separate filter systems and photographed serially or on one frame. Fluorescent antibody molecules are small enough not to mask each other's reactions if two (or more) antigens are localized on the same structure. The primary antibodies (which do not have to be from the same species) may be applied sequentially or together, provided that each is at its predetermined optimal concentration in the mixture and that the preparation method is suitable for all the antigens.

If enzyme-labelled antibodies are used, the enzymes must, of course, be developed separately. The antibodies can be applied one by one, each followed by development of its enzyme, or together, with the enzymes developed separately later. The enzyme end-products must be of different colours and the development medium for one must not damage the activity of the next to be developed. Different enzyme labels must be used if the labelled antibodies are applied simultaneously, for example, peroxidase, alkaline phosphatase and β-D-galactosidase, but with sequential application all the antibodies could be labelled with peroxidase. Application of these must be separated by peroxidase development to give different colours and repeat blocking with H_2O_2 to inactivate any residual peroxidase from the first labelled antibody before going on to the second peroxidase-labelled antibody and developing the enzyme with a second chromogen. Possible colours could be black (nickel-intensified DAB), brown (DAB), red (AEC) and blue (4-chloro-1-naphthol or phenyl tetrazolium method). If there is a possibility of co-localization of antigens, care must be taken that the end-product of the first to be developed does not mask the product of the second or third reaction (Valnes and Brandtzaeg, 1982).

9.2 Double immunostaining with primary antibodies raised in different species, or of different immunoglobulin sub-class (Appendix A.11.1)

9.2.1 Double immunoenzymatic method

This method (Mason and Sammons, 1978) is recommended because it avoids any possibility of cross-reactivity between the two antigen–antibody systems. The two primary antibodies, for example mouse anti-1 and rabbit anti-2, are applied together, each at its optimal dilution in the mixture. The second layer is a mixture of optimally diluted, species-specific, differently labelled anti-rabbit and anti-mouse immunoglobulins. The species in which the second layers are made must be non-cross-reactive, so that, for example, it is not possible with this method to use a rabbit anti-mouse immunoglobulin with a goat anti-rabbit immunoglobulin because the two would react together. If the second antibodies are unlabelled, a rabbit PAP can be combined with a mouse APAAP method in an adaptation of the original two-layer method (*Figure 9.1; Plates 6 and 12*).

It is also possible to use a biotin-labelled primary or secondary antibody for one of the reactions and an ABC or labelled avidin to detect it. It is possible to use two monoclonal primary antibodies if they are of different immunoglobulin classes (e.g. IgG and IgM) or sub-classes (e.g. IgG_1 and IgG_2) with class- or sub-class-specific secondary antibodies.

An essential control is to show that the second antibodies are truly species-specific and do not cross-react with the inappropriate primary under the conditions of the experiment.

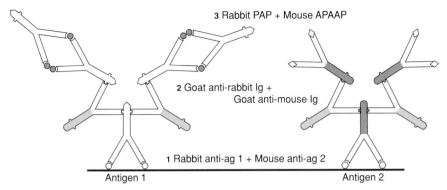

Figure 9.1: Simultaneous or sequential double staining by the double-immunoenzymatic method.

The original method described a peroxidase-labelled system in conjunction with an alkaline phosphatase-labelled system, the two enzymes being separately developed at the end of the two simultaneous reactions. It is not important which enzyme is developed first, but it is sometimes easier to find an appropriate end-point for the peroxidase/DAB/H_2O_2 reaction if the alkaline phosphatase has been developed first to its maximum with a blue end-product. This applies particularly where two antigens are suspected of being in the same structure, labelled by both immunoreactions, whereupon mixing of the blue and brown end-products occurs, giving a purple-brown. This mixed colour is easily distinguishable from the clear blue and brown of the separately developed enzymes, provided that the DAB reaction is stopped when a fairly light brown colour has developed. If the reaction is allowed to proceed to dark brown it may mask the lighter blue product. The AEC peroxidase reaction gives a red end-product which is also useful in double stains with a blue alkaline phosphatase product. The colour contrast is attractive and mixed colours show two antigens on the same site in purple. It is useful to do the double immunoenzymatic stain in triplicate, developing each enzyme separately as well as in combination, or with single antibodies as well as mixed, in order to compare and validate the results.

If the two antigens have different optimal preparation methods, such as a preference for protease digestion for one antigen only, or protease digestion being effective on one antigen but ineffective on the other, which requires heat-mediated retrieval, the two systems can be applied sequentially, the enzyme of the first being developed before the second pre-treatment is applied, followed by the second primary antibody and the rest of the second reaction.

The method is adaptable to many colour combinations depending on the enzyme labels (*Plate 12*, p. 38) (see Chapter 4). If either of the end-products is solvent-labile, a water-based mountant must be used.

9.2.2   Double immunofluorescence method

The non-cross-reactivity principles of the double immunoenzymatic method apply. Double immunofluorescence (see front cover) avoids problems of colour mixing since only one label is seen at a time. Co-localization is established by photomicrography or digital image capture.

9.2.3   Triple immunostaining

The number of antigens stained in a single preparation with this type of method is limited only by the number of primary antibodies available raised in different species or of different classes and their appropriate non-cross-reacting, specific second antibodies with distinguishable labels. However, it is possible to combine a method for double staining using primaries of the same species with double immunoenzymatic labelling as described above, provided that at least one of the antigens is known to be in a separate structure from the other two, which may or may not be separated.

   The known separate antigen is first immunostained by any method that will provide a very dark end-product, for example immunoperoxidase with DAB and nickel enhancement (see Section 8.2.2) or immunogold with silver intensification (see Section 4.3). The double immunoenzymatic method is then carried out for the other two antigens, and the intensity of the first reaction product should mask any cross-over that may occur, resulting in three differently stained antigens in the same preparation (*Plate 13(a)*, p. 39) (Van Noorden *et al.*, 1986).

9.3   Unlabelled primary antibodies from the same species

9.3.1   The problem

Frequently, when double labelling would be an advantage, only monoclonal mouse or only polyclonal rabbit antibodies are available. Using the simplest example of a double indirect method with enzyme-labelled second antibodies, it is obvious that if two primary antibodies from the same species are applied simultaneously they will each bind to their own antigens, but the labelled second layer will bind to both indiscriminately. Consider sequential application. One mouse monoclonal antibody is applied, followed by an enzyme-labelled anti-mouse Ig and developed. When the next mouse monoclonal antibody is applied it may find unused binding sites on the first anti-mouse Ig as well as its own antigen, and the next (differently labelled) anti-mouse Ig will bind to the second primary not only on its tissue site, but also to any attached to the first secondary antibody and to any unused binding sites on the first primary (*Figure 9.2*). After development of the enzyme on the second reaction, the result would be a pure colour on the site of the second primary antibody attachment and a mixed colour on the site of the first. If the two antigens are known to be in different locations, this might be acceptable, but if the localization is unknown, there would be no way of telling whether the mixed colour on the first site was due to dual localization of the antigens or to antibody cross-over. The same considerations would apply to double fluorescence methods.

Very many ingenious methods have been suggested for overcoming this problem. A few which work well are described here, but they are now mainly of historic interest as two more convenient methods (heat blocking and ARK biotinylation) have emerged which go a long way towards solving the problem.

9.3.2   Elution methods

Elution of immunoreactants leaving reaction product. The problem of using labelled primary antibodies is the relative insensitivity

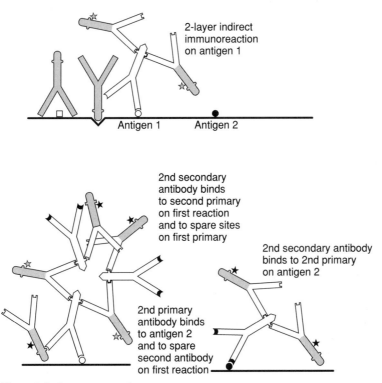

Figure 9.2: Crossover reactions.

of the method and the need to label each antibody. Various indirect methods have been suggested and here the main problem is to avoid cross-reactivity. Nakane (1968) achieved immunostaining of three hormones in the rat pituitary gland with rabbit primary antisera and indirect immunoperoxidase reactions with differently coloured end-products. He avoided the problem of the second reaction's primary rabbit antibody binding to the first reaction's anti-rabbit immunoglobulin and the second reaction's anti-rabbit immunoglobulin binding to the first reaction's primary by eluting the reagents of the first reaction after development in a prolonged exposure to acid buffer, leaving on the tissue the insoluble end-product, for example the brown precipitate from DAB. The second antibody was then applied and the rest of the method repeated using a different chromogen, such as 4-chloro-1-naphthol, to give a blue-grey coloured product. The elution procedure was then repeated and a third immunostain applied to give, for example, a red colour with AEC or, as used by Nakane (1968), α-naphthol with pyronin. The immunostaining could be done with the PAP method for greater sensitivity (*Figure 9.3*).

Very careful controls are necessary to ensure that all the first reaction antibodies have been eluted before the second set are applied (see Nakane, 1968, for details). Failure to remove all the antibodies may result in cross-reactivity, giving a false impression that antigens are co-localized because of mixing of end-products. Another problem of this method is that highly avid antibodies which are useful for immunocyto-chemistry are difficult to dislodge from their antigens, even in strong acid.

Elution of antibodies and dissolution of reaction product. Tramu *et al.* (1978) introduced a slightly different double-staining method, less damaging to the tissue and other antigens, whereby the first immunoperoxidase stain is labelled with 4-chloro-1-naphthol as chromogen. The section is then photographed and the antigen–antibody complex is dissociated by a short oxidation period in acidified potassium permanganate. The naphthol reaction product may then be dissolved in alcohol. Restaining, starting with the second antibody of the first reaction, should give a negative result, indicating that the complex has been entirely removed. The second stain is then carried out, with a different chromogen, and the section is photographed. Photographs of the same area may be compared. If the naphthol stain is not dissolved, the two antigens may be seen simultaneously.

More recently, Espada *et al.* (2001), using a diffferent eluting system that disrupted the immunoglobulin molecules, achieved multiple indirect immunofluorescence staining on a single-cell culture preparation by eluting the antibodies and fluorescent label with sodium dithionite and re-staining with a different primary antibody. The same second fluorescent antibody could be used.

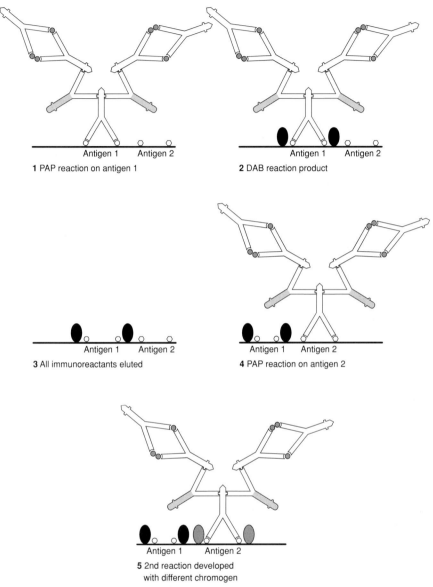

1 PAP reaction on antigen 1

2 DAB reaction product

3 All immunoreactants eluted

4 PAP reaction on antigen 2

5 2nd reaction developed
with different chromogen

Figure 9.3: Sequential double staining with elution of first antigen–antibody reactions.

9.3.3   Indirect double immunostaining without elution

Masking the first reaction site with reaction product. Sternberger and Joseph (1979) suggested that sequential double immunostaining with two primary antibodies raised in the same species can be carried out successfully without dissociation of the first reaction, provided that the first (immunoperoxidase) is developed with DAB to such an extent that the reaction product masks any unreacted binding sites on the first set of immunoglobulins. The second primary will then bind only to its own antigen. This method is satisfactory provided it is known that the two antigens are located in different structures. If both were in the cytoplasm of the same cell, the reaction product of the first reaction would prevent the second primary antibody from finding its target.

It may be difficult to be certain that the peroxidase development of the first reaction is adequate, and if it is not, cross-over will occur and a mixed colour will be obtained at the site of the first reaction after development of the second peroxidase in a different colour. A way of avoiding this is to make the first reaction product very dark, with the cobalt or nickel modification of the DAB reaction or an immunogold–silver reaction for the first of the pair, and to develop the second reaction in a lighter colour. Any crossing over in this case will be undetectable (*Plates 4* and *13(a)*).

Deactivation of potentially cross-reacting groups with hot formaldehyde vapour. Wang and Larsson (1985) carried out double direct immunofluorescence with sequential immunoreactions using fluorescein- and rhodamine-labelled anti-rabbit immunoglobulins as second layer antibodies. Between the two reactions, the preparation was rinsed in distilled water, dried and exposed for several hours to formaldehyde vapour from paraformaldehyde at 60–80°C. This treatment blocked spare combining sites on both the primary and secondary antibodies of the first reaction, while retaining the fluorescence. The second reaction was then carried out. Enzyme-labelled antibodies could be used instead of fluorescent ones. The principle also applied to double labelling with immunogold at an ultrastructural level (see Chapter 10).

Deactivation of potentially cross-reacting groups with heat. Wang and Larsson (1985) found that in some cases, heat alone was sufficient to block the reactions. The same effect was found by Kolodziejczyk and Baertschi (1986) who used a higher temperature (130°C) for a shorter time (4 minutes).

A reliable multiple staining method using heat inactivation was suggested by Lan *et al.*, (1995). They exploited the microwave oven then coming into use for antigen retrieval, and showed that after applying an enzyme-labelled antibody and developing the enzyme, microwaving for 10 minutes in 0.01 M citrate buffer blocked completely any immunoglobulin-binding sites on the first reaction. A second immunoenzyme reaction could then be applied, with a primary antibody from the same or a different species, and the process could be repeated for a third reaction, and even a fourth. This method can be used with sensitive techniques such as PAP, APAAP and ABC methods. Residual alkaline phosphatase is destroyed by the microwave step, while the end-product of reaction stays in the preparation. It is thus possible to use alkaline phosphatase developed with different chromogens as the sole label in multiple staining (Deininger and Meyermann, 1998). Residual peroxidase is probably inhibited, but a hydrogen peroxide block between layers is advisable if peroxidase is one of the labels. If avidin–biotin systems are being used, extra biotin may be revealed by the microwave step, so a biotin block between layers is also necessary.

Although applied immunoglobulin binding sites are blocked by the heat method, endogenous immunoglobulins may be further exposed. It is therefore possible that unwanted tissue binding may occur with the second (and subsequent) immunoreactions. If this occurs, a serum blocking step should be incorporated before the antibodies are applied. A sample method is given in the Appendix (A.11.2) but readers will be able to work out methods according to their own reagents.

The number of reactions that can be carried out sequentially is limited only by the survival power of the preparation under repeated microwaving and the availability of suitable labels, provided that the antigens are localized in different sites. Mixed colour reactions are suitable only for pairs of antigens. Unlike the ARK method described below, this method allows the reliable use of polyclonal antibodies from the same species in multiple staining, which is often useful (*Plates 13(b)* and *14*).

Of course numerous negative controls are required to show that there is no cross-reactivity between layers. The important controls for this method are to ensure that the second antigen survives the blocking treatment and, by applying an irrelevant rabbit antibody or normal rabbit serum and the second labelled antibody, to ensure that all potential binding sites have been blocked. An additional safety precaution is to reverse the order of reaction, starting with the second primary antibody, to ensure that the original first reaction has not blocked access to the second antigen. In general, it is best to carry out the weaker reaction first.

Immunological blocking of potentially cross-reacting groups.
Lewis-Carl *et al.* (1993) used inappropriate immunoglobulin from the same species as the primary antibody to occupy any remaining antigen-combining sites on the second (fluorescent) antibody of the first reaction. They then blocked all antibody-binding sites on the blocking immunoglobulin and any remaining on the primary antibody by using F(ab) fractions of unlabelled second antibody. It was necessary to use F(ab) fractions and not F(ab)$_2$ fractions to ensure that no antigen-binding sites remained free. The second primary antibody was then applied, and could not bind to the first secondary because all free sites were occupied by inappropriate antibody of the same species, but could only bind to the second antigen. The second fluorescent antibody could not bind to the first primary since any spare sites were occupied by F(ab) fractions of immunoglobulin from unlabelled second antibody, and could only bind to the second primary.

In theory, this is a trouble-free way of ensuring double immunofluorescence with two primary antibodies from the same species. In our experience, following this method with enzyme-lablelled antibodies, it is better to use monoclonal than polyclonal primaries in order to minimize the types of immunoglobulin binding site that must be blocked by the Fab fractions, and it is difficult to get the reagents to the right concentration for complete blocking unless sub-optimal levels of primary antibody are used.

Combining primary and labelled secondary antibodies* in vitro *before immunostaining. Another ingenious method for overcoming the problem of cross-over reactions is that of Krenács *et al.* (1991). This requires each primary antibody to be mixed in various proportions (up to 30 combinations) with its labelled secondary antibody. After incubation, normal serum from the species providing the primary antibody is added to occupy any unused sites on the secondary antibody. The complexes are tested separately and those giving the best staining are applied to the tissue simultaneously or sequentially, resulting in double labelling.

Double immunostaining using the Dako ARK kit. The principle of combining primary and labelled secondary antibodies *in vitro* was used by Dako to produce their Animal Research Kit (ARK) for staining with mouse antibodies on mouse tissue (see Section 7.2.3). van der Loos and Göbel (2000) realised that the kit could be used in a double stain with two monoclonal mouse antibodies, combined with immunological blocking of residual cross-reactivity. In their method, the first immunoreaction is carried out normally, but the enzyme is not developed. Any residual anti-mouse Ig binding sites are blocked with normal mouse Ig, then the second monoclonal antibody, directly labelled with F(ab) fractions of biotinylated goat anti-mouse Ig is applied and developed with labelled avidin. The alkaline phosphatase label is then developed, followed by the peroxidase (*Plate 15*). The method is simplest if a non-biotin–avidin method is used for the first antibody (PAP or APAAP or EnVision, if a sensitive method is needed), but if an avidin–biotin method is necessary, then a biotin blocking step must be done between the layers after the normal serum block.

There seems to be no reason why the reaction could not be repeated, after biotin blocking, with a third ARK-biotinylated antibody and another label.

This chapter has described but a few of the very many methods that have been put forward for multiple immunostaining. The principles outlined should help the reader to devise suitable techniques for new applications.

References

Cordell JL, Falini B, Erber WN, Ghosh AK, Abdulaziz Z, MacDonald S, Pulford KAF, Stein H, Mason DY. (1984) Immunoenzymatic labeling of monoclonal antibodies using immune complexes of alkaline phosphatase and monoclonal anti-alkaline phosphatase (APAAP comlexes). *J. Histochem. Cytochem.* **32**, 219–222.

Deininger MH, Meyermann R. (1998) Multiple epitope labelling by the exclusive use of alkaline phosphatase conjugates in immunohistochemistry. *Histochem. Cell Biol.* **110**, 425–430.

Espada J, Juarranz A, Villanueva A, Cañete M, Andrés I, Stockert J. (2001) Recycling cultured cells for immunofluorescent labeling. *Histochem. Cell Biol.* **116**, 41–47.

Kolodziejczyk E, Baertschi AJ. (1986) Multiple immunolabeling in histology: a new method using thermal inactivation of immunoglobulins. *J. Histochem. Cytochem.* **34**, 1725–1729.

Krenács T, Hirotugu U, Tanaka S. (1991) One-step double immunolabeling of mouse interdigitating reticular cells: simultaneous application of pre-formed complexes of monoclonal rat antibody M1-8 with horseradish peroxidase-linked anti-rat immunoglobulins and of mouse monoclonal anti-Ia antibody with alkaline phosphatase-coupled anti-mouse immunoglobulins. *J. Histochem. Cytochem.* **39**, 1719–1723.

Lan HY, Mu W, Nikolic-Paterson DJ, Atkins RC. (1995) A novel, simple, reliable and sensitive method for multiple immunoenzyme staining: use of microwave oven heating to block antibody crossreactivity and retrieve antigens. *J. Histochem. Cytochem.* **43**, 97–102.

Lewis-Carl S, Gillete-Ferguson I, Ferguson DG. (1993) An indirect immunofluorescence procedure for staining the same cryosection with two mouse monoclonal antibodies. *J. Histochem. Cytochem.* **41**, 1273–1278.

Mason DY. (1985) Immunocytochemical labeling of monoclonal antibodies by the APAAP immunoalkaline phosphatase technique. In *Techniques in Immunocytochemistry*, Vol. 3 (eds GR Bullock, P. Petrusz). Academic Press, New York, pp. 25–42.

Mason DY, Sammons RE. (1978) Alkaline phosphatase and peroxidase for double immunoenzymatic labelling of cellular constituents. *J. Clin. Pathol.* **31**, 454–462.

Nakane PK. (1968) Simultaneous localization of multiple tissue antigens using the peroxidase-labeled antibody method: a study on pituitary gland of the rat. *J. Histochem. Cytochem.* **16**, 557–560.

Sternberger LA, Joseph FA. (1979) The unlabeled antibody method. Contrasting color staining of paired pituitary hormones without antibody removal. *J. Histochem. Cytochem.* **29**, 1424–1429.

Tramu G, Pillez A, Leonardelli J. (1978) An efficient method of antibody elution for the successive or simultaneous localization of two antigens by immunocytochemistry. *J. Histochem. Cytochem.* **26**, 322–324.

Valnes K, Brandtzaeg P. (1982) Comparison of paired immunofluorescence and paired immunoenzyme staining methods based on primary antisera from the same species. *J. Histochem. Cytochem.* **30**, 518–524.

van der Loos CM. (1999) *Immunoenzyme Multiple Staining Methods.* BIOS Scientific Publishers, Oxford.

van der Loos CM, Göbel H. (2000) The Animal Research Kit (ARK) can be used in a multistep double staining method for human tissue specimens. *J. Histochem. Cytochem.* **48**, 1431–1437.

Van Noorden S, Stuart MC, Cheung A, Adams EF, Polak JM. (1986) Localization of pituitary hormones by multiple immunoenzyme staining procedures using monoclonal and polyclonal antibodies. *J. Histochem. Cytochem.* **34**, 287–292.

Wang BL, Larsson L-I. (1985) Simultaneous demonstration of multiple antigens by indirect immunofluorescence or immunogold staining. Novel light and electron microscopical double and triple staining method employing primary antibodies from the same species. *Histochemistry* **83**, 47–56.

10 Immunocytochemistry for the Transmission Electron Microscope

This chapter cannot cover the wide field indicated by its title, so it will be limited to standard methods for sections cut from embedded blocks and mounted on grids. For further information see the work by Monaghan *et al.* (1993), Polak and Priestley (1992) and Skepper (2000).

10.1 Principles

The principles of immunocytochemistry for the electron microscope are exactly the same as for the light microscope. The immunoreactivity of the antigen must be preserved in a fixed tissue context and the immuno-reaction must be specific and efficient, with no background labelling. The differences are that the label must be electron-dense and the embedding material is a resin which can be sectioned very thinly. Fixed frozen sections can also be used, but this application is rather specialized. All the standard controls apply as in light microscopical methods.

10.2 Fixation

Fixation must be started as soon as possible after removal of the tissue sample. Perfusion fixation is recommended, if possible, as it allows the fixative to penetrate the tissue thoroughly and rapidly. Tissue samples for immersion fixation must be small (less than 1 mm^3) for the same reason, and to avoid damage to the antigen, fixation times should be kept short (30–120 min). Suspended cultured cells can be centrifuged then held together in agar gel (4%) for fixation and processing.

Useful fixatives for ultrastructural immunocytochemistry include weak mixtures of paraformaldehyde and glutaraldehyde, such as 2% paraformaldehyde/0.05% glutaraldehyde (Monaghan *et al.*, 1993). Other combinations have also been used. The paraformaldehyde penetrates fast, but fixes slowly, while the glutaraldehyde component penetrates the tissue slowly, but fixes rapidly; the combination of the two allows rapid penetration and fixation. PLP (see Section 3.1.3) can also be useful. For some antigens, post-osmication of the tissue sample can be used, giving better tissue preservation (particularly of lipid membranes) without damaging the immunoreactivity. In addition, the fixation effect of osmium is partially reversible by treating the sections with an oxidizing agent such as sodium metaperiodate (Bendayan and Zollinger, 1983). However, the preparation of un-osmicated samples is always necessary when testing an unknown antigen–antibody reaction and, as for light microscopical immunocytochemistry, it may be necessary to try a number of different fixatives to find the best. It is often useful to test the methods at light microscopical level first, on sections from paraffin-embedded or frozen blocks fixed in a variety of ways suitable for electron microscopical immunocytochemistry, or on semi-thin sections from resin-embedded blocks that can subsequently be used at electron microscopical level.

10.2.1   Pre-embedding immunocytochemistry

This procedure, with peroxidase as the label, is suitable for thick sections of pre-fixed frozen tissue, Vibratome sections, cell or tissue culture preparations or membranes such as iris diaphragms. Areas are then selected under the light microscope, processed and embedded in resin for ultrathin sectioning. This method is limited by the small distance that antibodies can penetrate into the tissue, and immunoperoxidase is used because gold-labelled antibodies penetrate very poorly (Priestley *et al.*, 1992). However, ultra-small gold labels such as nanogold (see Section 4.3.2) can improve the penetrability of reagents considerably (Sawada and Esaki, 2000). In ultrathin sections, visualization of nanogold, as of 1 nm diameter colloidal gold, is difficult, but silver enhancement on the grid overcomes the difficulty. It builds up on the gold particles, giving good contrast and making the reaction very sensitive. If cell surface antigens are to be localized on cells in suspension, the primary antibody and even the labelled secondary antibody can be applied before the cells are fixed, as for light microscopical immunocytochemistry. This allows conventional glutaraldehyde fixation to be used, followed by osmium tetroxide, providing good morphology (*Figure 10.1*).

10.2.2 Non-embedding immunocytochemistry

This specialised technique avoids resin embedding by using fixed, cryo-protected, snap-frozen tissue from which ultrathin sections are cut at very low temperature and mounted on grids for immunostaining.

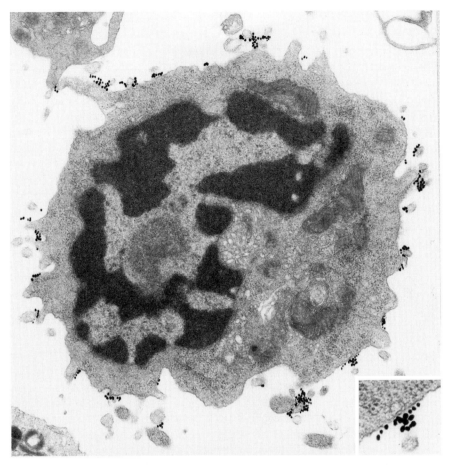

Figure 10.1: Normal human T-lymphocyte. Electron micrograph showing cell membrane labelled with mouse monoclonal anti-human CD 3 (T-cell marker). Localization was with 30 nm diameter colloidal gold-labelled goat anti-mouse IgG. F(ab)$_2$ fragments of the antibodies were used to avoid the antibodies binding to Fc receptors. Immunostaining was carried out on unfixed cells in suspension and was followed by glutaraldehyde fixation and standard processing. Ultrathin immunolabelled sections were contrasted with lead citrate and uranyl acetate. Original magnification: × 15 000. Inset: higher magnification of a group of colloidal gold particles at an immunoreactive site on the cell membrane. Courtesy of Dr. E. Matutes, Department of Haematology, Royal Marsden Hospital, London.

10.3   Processing to resin

Several kinds of resin are suitable for embedding. Traditionally, epoxy resins such as araldite or epon have been used. These require complete dehydration of the tissue through graded alcohols, followed by infiltration with propylene oxide then a mixture of propylene oxide and resin and, finally, the resin itself and polymerization at 40–60°C. If semi-thin (1 μm) sections mounted on slides are to be immunostained the resin must be removed from the sections, using, for example, a saturated solution of sodium hydroxide in alcohol (Lane and Europa, 1965) (Appendix, Section A.9.1).

Among the acrylic resins, LR White or Lowicryl K4M or HM20 have been used extensively for ultrastructural immunocytochemistry. They are tolerant of some water in the tissue, and Lowicryl can be polymerized by ultraviolet light at very low temperatures, which may be useful for heat-labile antigens. Lowicryl resins can be introduced into unfixed tissue after freeze-substitution or into lightly fixed tissue cooled by progressive lowering of temperature of graded alcohols. Semi-thin sections from acrylic resin blocks can be stained without removal of the resin, but because of the hydrophilic nature of the resins, special care must be taken to keep them on the slides, using a silane coating and overnight drying at 37°C.

10.4   Labels

Ferritin was one of the first labels to be used for immunolabelling at the electron microscope level, because of the distinctive shape of the crystals and their electron density. However, it was soon supplanted for most purposes by peroxidase, using the DAB reaction product which becomes electron-dense after osmication (*Figure 10.2*). Problems with peroxidase as a label concern the DAB reaction product which can be so dense that it obscures the structure of the underlying organelle and has a tendency to creep along membranes. This is not usually noticeable at the light microscopical level but could be critical for ultrastructural antigen localization. Colloidal gold (Faulk and Taylor, 1971) has superseded peroxidase because of its extreme electron density, which permits conventional contrasting of the section with lead citrate and uranyl acetate, and its particulate nature which allows the underlying structure to be seen (*Figure 10.3*). In addition, the particles can be made with selected diameters from 1 to 40 nm, so that immunogold methods lend themselves to multiple labelling. In general, the smallest size of gold particle compatible with easy detection should be used, to permit as much labelled antibody as possible to reach the antigenic sites. The ultrasmall nanogold label is useful in this respect. For a full discussion and practical details of all aspects of gold and silver labelling, see Hacker and Gu (2002). The gold can be enhanced with silver, as for light microscopy, which can be useful for expanding the size of very small particles and for differentiating between two sequential labelling procedures on the same specimen.

Labelling of antibodies with colloidal gold can be done quite easily in the laboratory (De Mey, 1986b), but many types of gold-labelled reagent

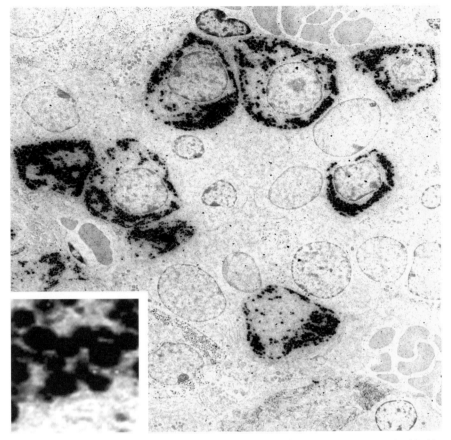

Figure 10.2: Rat interior pituitary gland fixed in glutaraldehyde, no osmication, embedded in araldite. Electron micrograph of ultrathin section immunostained for growth hormone by an indirect immunoperoxidase technique. Peroxidase was developed with H_2O_2 and DAB as the chromogen, followed by osmium tetroxide to make the reaction product electron-dense. No further contrast as applied. The positive cells are easily visible at low magnification. At high magnification (inset) the structure of the secretory granules is obscured by the reaction product (see *Figure 10.3*). Courtesy of Dr. C. Sarraf, Imperial College Faculty of Medicine, Hammersmith Hospital, London.

are available commercially. The diluent for gold-labelled reagent must contain a high concentration of protecting protein (0.05 M Tris/HCl buffer, pH 8.2, containing 1% bovine serum albumin, BSA). Gold-labelled immunoglobulins should be centrifuged (2000 *g* for 10 min) after optimal dilution to remove any micro-aggregates of gold particles.

10.5   Sectioning resin blocks

Sections are cut at 60–100 nm and picked up on uncoated nickel or gold grids. They can be stored indefinitely.

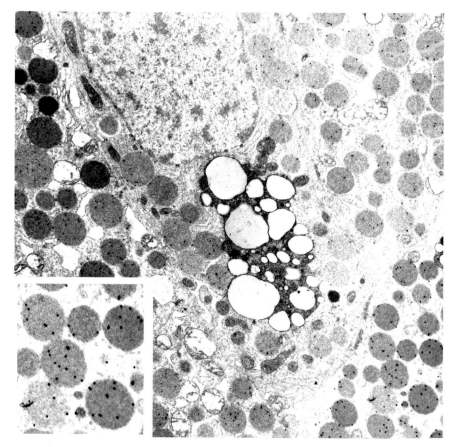

Figure 10.3: Somatostatin-secreting tumour, fixed in glutaraldehyde followed by osmium tetroxide and embedded in araldite. Electron micrograph of ultrathin section pretreated with sodium metaperiodate then immunostained with a rabbit polyclonal antibody to somatostatin followed by 10 nm diameter colloidal gold-adsorbed goat anti-rabbit IgG. The section was then contrasted with lead citrate and uranyl acetate. Inset: higher magnification of the immunola-belled secretory granules. Note the clarity of the labelling compared with *Figure 10.2*, but these labelled cells would be harder to identify at low magnification.

10.6   Pre-treatment

Epoxy resin sections of non-osmicated material may be etched with 10% H_2O_2 to give better contact of the antibody with the section. For osmicated sections, saturated sodium metaperiodate treatment for 10–30 min provides both etching and some reversal of the fixation effects of osmium (Bendayan and Zollinger, 1983). After etching, the grids should be washed well in micropore-filtered distilled water before proceeding to the immunostaining protocol (see Appendix, Section A.12).

It is not usually necessary to apply antigen retrieval methods to sections on grids, but, as for light-microscopical immunocytochemistry, heat mediated antigen retrieval is becoming increasingly used as a way of increasing the sensitivity of immunolabelling on aldehyde-fixed material. There are many examples in the literature (Sormunen and Leong, 1998; Groos *et al.*, 2001; Hann *et al.*, 2001). The mechanism of retrieval may not be identical to the supposed unmasking in paraffin sections by breaking intra-protein fixative bonds, but may be more concerned with effects on the resin. For reviews and discussion, see Brorson (2001) and Mayer and Bendayan (2001).

10.7   Immunolabelling procedure (Appendix A.12)

Grids are floated on drops of the various buffer and antibody solutions, either in 15 μl multiwell Terasaki plates or on strips of dental wax or Parafilm or poured paraffin wax in a Petri dish. They are transferred from drop to drop with grid forceps, draining excess liquid between drops by touching the rim of the grid to the filter paper.

The methods used are the same as for light microscopical immunocytochemistry, comprising the directly labelled, indirect and three-layer methods. One additional method is available which uses Protein A, a bacterial protein from the coat of *Staphylococcus aureus* which can be adsorbed to colloidal gold in the same way as an antibody and has the property of binding to the Fc portion of many types of immunoglobulin. It can thus be used as a universal second layer with little worry about species – though a similar protein, Protein G, from the cell walls of *Streptococcus* binds better than Protein A to immunoglobulins from rat and goat and to some sub-classes of human and mouse immunoglobulin (Bendayan and Garzon, 1988). Background blocking should be with albumin or another inert protein rather than with serum if these proteins are to be used, as they may otherwise bind to the immunoglobulins in the blocking serum. If direct, indirect or three-layer techniques are used, blocking can be with the appropriate normal serum or with albumin (usually 1% BSA, which should be globulin-free). Tris buffer or PBS can be used, and antibody dilutions are usually about the same as for light microscopy, in buffer containing 0.1% BSA and (if solutions are to be stored) 0.1% sodium azide.

10.7.1   Immunolabelling with peroxidase

The steps are the same as for light microscopical immunocytochemistry. Peroxidase development is followed by osmication with 1% osmium tetroxide to make the DAB reaction product electron dense.

Safety note: Osmium tetroxide vapour is harmful. Use a fume cupboard.

10.7.2   Amplification

The amount of label on a section can be enhanced in the same ways as for light microscopy, with repetition of layers and the deposition of metallic silver. The tyramine amplification system can also be used (for review see Mayer and Bendayan, 2001). For example, a biotinylated antibody is followed by peroxidase-labelled streptavidin, then biotinylated tyramine with hydrogen peroxide, followed by a second application of streptavidin labelled with gold.

10.8   Contrasting

At the end of the reaction, gold-labelled sections may be contrasted with lead citrate and uranyl acetate in the conventional way. Peroxidase-labelled sections may be so contrasted if desired, but it is useful to keep an uncontrasted preparation for comparison in case the end-product is not sufficiently dense to be seen against the background structure.

10.9   Multiple labelling

Because colloidal gold particles can be made in different sizes, simultaneous direct double labelling with differently labelled antibodies (any species) or indirect labelling with primary antibodies raised in different species and non-cross-reacting, second antibodies labelled with gold particles of different sizes are very useful techniques (*Figure 10.4*) (Tapia *et al.*, 1983). In indirect double labelling, the same stringent controls and precautions against cross-reaction must be observed as for light microscopical double labelling. Gold-labelled Protein A provides an alternative to antibody labelling and can be used for double labelling with primary antibodies raised in the same species, though care must be taken to block unwanted binding sites between the two applications (Roth, 1982). On-grid immunolabelling offers a further way of double labelling by using one side of the grid for one (or two) localization(s), and the reverse side for a sequential second (or third). If two of the antibodies are raised in the same species, this is a good way of avoiding cross-reaction, but great care must be taken to keep each set of reagents to its own side of the grid (Bendayan, 1982).

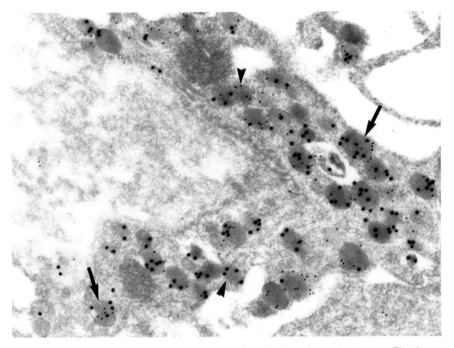

Figure 10.4: Human pituitary tumour secreting both prolactin and growth hormone. The tissue was fixed in paraformaldehyde and embedded in araldite. Electron micrograph of an ultrathin section showing simultaneous double immunogold labelling. The primary antibody layer was a mixture of rabbit anti-growth hormone and monoclonal mouse anti-prolactin, followed by a mixture of 10 nm diameter colloidal gold-adsorbed goat anti-rabbit IgG and 5 nm diameter colloidal gold-adsorbed goat anti-mouse IgG. The section was contrasted with lead citrate and uranyl acetate after immunolabelling. Most secretory granules contain both hormones (arrows): some contain only one of the two (arrowheads).

ARK-biotinylated monoclonal antibodies could be used in double labelling in the same way as for light microscopy (Section 9.3), the detecting streptavidin being labelled with colloidal gold particles of a different diameter from those of the preceding layer. With appropriate blocking between layers, multiple labelling strategies could be devised.

References

Bendayan M. (1982) Double immunocytochemical labeling applying the protein A–gold technique. *J. Histochem. Cytochem.* **30**, 81–85.

Bendayan M, Garzon S. (1988) Protein G–gold complex: comparative evaluation with Protein A–gold for high resolution immunocytochemistry. *J. Histochem. Cytochem.* **36**, 597–607.

Bendayan M, Zollinger M. (1983) Ultrastructural localization of antigenic sites on osmic-fixed tissue applying the Protein A–gold technique. *J. Histochem. Cytochem.* **31**, 101–109.

Brorson SH. (2001) Heat-induced antigen retrieval of epoxy sections for electron microscopy. *Histol. Histopathol.* **16**, 923–930.

De Mey J. (1986b) The preparation and use of gold probes. In *Immunocytochemistry, Modern Methods and Applications* (eds JM Polak, S Van Noorden). Butterworth–Heinemann, Oxford, pp. 115–145.

Faulk WR, Taylor GM. (1971) An immunocolloid method for the electron microscope. *Immunochemistry* **8**, 1081–1083.

Groos S, Reale E, Luciano L. (2001) Re-evaluation of epoxy resin sections for light and electron microscopic immunostaining. *J. Histochem. Cytochem.* **49**, 397–406.

Hacker GW, Gu J. (eds) (2002) *Gold and Silver Staining: Techniques in Molecular Morphology.* CRC Press, Boca Raton.

Hann CR, Springett MJ, Johnson DH. (2001) Antigen retrieval of basement membrane proteins from archival eye tissues. *J. Histochem. Cytochem.* **49**, 475–482.

Lane BP, Europa DL. (1965) Differential staining of ultrathin sections of epon-embedded tissue for light microscopy. *J. Histochem. Cytochem.* **13**, 579–582.

Mayer G, Bendayan M. (2001) Amplification methods for the immunolocalization of rare molecules in cells and tissues. *Prog. Histochem. Cytochem.* **36**, 3–85.

Monaghan P, Robertson D, Beesley JE. (1993) Immunolabelling techniques for electron microscopy. In *Immunocytochemistry, A Practical Approach* (ed JE Beesley). Oxford University Press, Oxford, pp. 43–76.

Polak JM, Priestley, JV. (eds) (1992) *Electron Microscopic Immunocytochemistry: Principles and Practice.* Oxford University Press, Oxford.

Priestley JV, Alvarez FJ, Averill S. (1992) Pre-embedding electron microscopic immunocytochemistry. In *Electron Microscopic Immunocytochemistry: Principles and Practice* (eds JM Polak, JV Priestley). Oxford University Press, Oxford, pp. 89–121.

Roth J. (1982) The preparation of Protein A–gold complexes with 3 nm and 15 nm gold particles and their use in labelling multiple antigens on ultrathin sections. *Histochem. J.* **14**, 791–801.

Sawada H, Esaki M. (2000) A practical technique to postfix nanogold-immunolabeled specimens with osmium and to embed them in Epon for electron microscopy. *J. Histochem. Cytochem.* **48**, 493–498.

Skepper JN. (2000) Immunocytochemical strategies for electron microscopy: choice or compromise. *J. Microsc.* **199**, 1–36.

Sormunen R, Leong AS-Y. (1998) Microwave-stimulated antigen retrieval for immunohistology and immunoelectron microscopy of resin-embedded sections. *Appl. Imunohistochem.* **6**, 234–237.

Tapia FJ, Varndell IM, Probert L, De Mey J, Polak JM. (1983) Double immunogold staining method for the simultaneous ultrastructural localisation of regulatory peptides. *J. Histochem. Cytochem.* **31**, 977–981.

11 *In Vitro* Methods for Testing Antigen– Antibody Reactions

The methods described below are not strictly immunocytochemical since they do not identify antigens *in situ,* but they can be useful to indicate whether serum from an immunized animal or clones from a hybridoma culture have the required immunoreactivity, and whether there is any cross-reactivity with other antigens. In some cases these assays can also be used to identify and quantify antigens in tissue extracts, offering a far more accurate quantification of the concentration of antigen in a sample than any immunocytochemical method, but lacking the advantage of localization. Where an immunocytochemical stain could pin-point one immunoreactive cell in a tissue section, the available antigen in an extract of the same tissue might be so diluted that the quantitative assays, despite their greater sensitivity, could not detect it.

It must be remembered that antigens *in vitro* may behave differently from the way they do in tissue where they are usually subjected to fixation and may be sequestered in cell organelles or attached to membranes. In the tissue context, only part of the antigen molecule may be visible to the applied antibody, and it may not be the part to which the antibody is most reactive, whereas in a radioimmunoassay the whole molecule is free to react, although it may be radiolabelled. In ELISA and blotting techniques the antigen is bound to a solid substrate, but is probably more easily approached by the antibody than in fixed tissue. These considerations explain why some antibodies are very useful in radioimmunoassay or Western blotting but very poor in immunocytochemistry. The reverse may also be true as regards Western blotting since, in this technique, the antigens are denatured by sodium dodecyl sulphate (SDS) before reaction, which might either reveal or destroy the particular epitope recognized by the antibody. Another cause for discrepancy is that in some *in vitro* tests pure antigen is offered to the antibody, whereas in tissue there are numerous other antigens which may react with the (polyclonal) antibody and which were not tested for *in vitro* because unknown. Antibodies for use in immunocytochemistry must be tested by immunocytochemistry.

Detailed discussion and practical details of these tests are beyond the scope of this book; the reader is referred to the work of Johnstone and Thorpe (1996) and Masseyef *et al.* (1993).

11.1   Radioimmunoassay

This method was proposed by Yalow and Berson (1959) and has become indispensable in clinical investigations of the concentration of hormones and other substances in blood and tissue extracts (for an introduction to the method see the work of Self *et al.*, 1976). It depends on the competition between a given quantity of a labelled antigen and the unlabelled antigen in the sample for a given quantity of specific antibody. The ratio between the bound and unbound antigen in the final solution gives a measure of the quantity present in the original sample. Note that here it is the antigen and not the antibody than carries the label, and that the label is a radioactive element, usually iodine. The method can be used for testing antibodies. It is so sensitive that useful antibody dilutions are usually several thousand times greater than for immunocytochemistry. However, this method can only be used as a rough guide as to whether the antibody is useful for immunocytochemistry, for the reasons mentioned above. If an antibody to human growth hormone, for example, is shown by radioimmunoassay to have as little as 0.001% binding capacity for prolactin, with which growth hormone shares some amino acid sequences, it may still show binding to prolactin in the immunocytochemical context where the concentration of antibody will be much greater than for radioimmunoassay and even a small population of cross-reacting antibodies will find their target.

11.2   Enzyme-linked immunosorbent assay (ELISA)

This method of testing for antibodies or antigens (Voller *et al.*, 1976; O'Beirne and Cooper, 1979) is nearer to immunocytochemistry than is radioimmunoassay as it uses built-up layers of antigens and antibodies and, usually, an enzyme label with a coloured end-product. However, the product is soluble and the density of colour gives a quantitative result.

When used for testing antibodies, say the first bleeding from a batch of immunized rabbits, a known quantity of the antigen is applied to the inside surface of the cups of a microhaemagglutination plate. The second layer consists of a series of dilutions of the antiserum to be tested, and the third layer is a standard preparation of alkaline phosphatase-conjugated goat anti-rabbit immunoglobulin. The enzyme is then developed and the colour in each cup can be estimated by eye or read in a colorimeter. Positive and negative controls are, of course, included to provide a standard for each test. The second antibody may be radiolabelled and the results read in a scintillation counter.

ELISA is less sensitive than radioimmunoassay, requiring somewhat lower dilutions of antibody approaching those used in immunocytochemistry, and is less reliable than radioimmunoassay for quantification since the amount of antigen attached to the plastic cup and the type of binding site left free are unpredictable. However, it more nearly resembles immunocytochemistry since it is the antibody and not the antigen that carries the label. ELISA provides a good screening test for antibodies and can be used to check for cross-reactivity or contaminating antibodies by lining the cups with a variety of antigens. These might include antigens known to be related structurally to the antigen under investigation, and also any carrier protein used in immunization. It could also be adapted to test for antigens in tissue extracts.

11.3   Western blotting

To identify exactly the tissue proteins with which an antibody is reacting, an extract of the tissue is subjected to electrophoresis on acrylamide gel. Because it is difficult to immunostain directly in the gel, the separated proteins are transferred to nitrocellulose paper by an electroblotting technique and the paper strip is then stained by the antibody, using any suitable detecting technique, as if it were a section of tissue, except that the strip is immersed in the antibody solutions. A parallel strip is stained to reveal the protein bands and matched up against the immunostained strip. The position of the protein is an indication of its molecular weight. A single, well-defined band indicates that an antiserum is specific for one antigen only, whereas multiple bands indicate several populations of antibodies. The identity of the antigen and the specificity and cross-reactivity of the antibody can be tested against blots of the relevant pure antigen and related or unrelated molecules.

11.4   Dot blots

An extremely simple way of testing antibodies against pure antigens is to apply very small quantities of antigens to a strip of filter paper, fix them with formaldehyde vapour and immunostain them in the same way as a tissue section (Larsson, 1981).

References

Johnstone A, Thorpe R. (1996) *Immunochemistry in Practice*, 3rd Edn. Blackwell Scientific Publications, Oxford.

Larsson L-I. (1981) A novel immunocytochemical model system for specificity and sensitivity screening of antisera against multiple antigens. *J. Histochem. Cytochem.* **29**, 408–410.

Masseyeff RF, Albert WH, Staines NA. (eds) (1993) *Methods of Immunological Analysis*. VCH-Verlagsgesellschaft, Weinheim.

O'Beirne AJ, Cooper HR. (1979) Heterogeneous enzyme immunoassay. *J. Histochem. Cytochem.* **27**, 1148–1162.

Self M, Rees LH, Landon J. (1976) An introduction to radioimmunoassay. *Med. Lab. Sci.*, 221–228.

Voller A, Bidwell D, Bartlett A. (1976) Microplate enzyme immunoassays for the immunodiagnosis of viral infections. In *Manual of Clinical Immunology* (eds NR Rose and H Friedman). American Society for Microbiology, Washington, DC, Ch. 69.

Yalow RS, Berson SA. (1959) Assay of plasma insulin in human subjects by immunological methods. *Nature* **184**, 1648–1649.

12   Applications of Immunocytochemistry

Immunocytochemistry is now one of the most widely used tools for research and diagnosis. Provided that clean antibodies of high specificity and avidity are available, the number of problems to which the method is applicable is unlimited. A few examples are given below.

12.1   Histopathological diagnosis

The introduction of antigen retrieval methods has meant that formalin-fixed, paraffin-embedded tissue can be used for most immunocytochemical staining, which has become a routine technique in histopathological diagnosis, and also in cytopathology, haematopathology, immunology and microbiology. Cytopathological specimens are usually centrifuged to make cytospin preparations for immunostaining and are fixed in alcohol. Haematological specimens (blood or bone marrow) are used as smears and fixed in acetone or acetone/methanol.

One of the main diagnostic uses is to determine the nature of tumours. Tumours often retain the antigenic characteristics of their tissue of origin, even when the cellular architecture is very undifferentiated, so immunostaining a biopsy from a metastasis can provide information about the probable site of the primary tumour (*Figure 12.1*). Cells that appear spindle-shaped, for example, could be shown to be of neural or muscular origin by immunostaining for the appropriate markers of nerves and muscles. A poorly differentiated tumour might be shown to be of lymphoid or epithelial nature by staining for cytokeratin and leukocyte common antigen (LCA). Cytokeratin is an intermediate filament, characteristic of epithelial cells and therefore a marker of carcinoma, a tumour derived from such cells. LCA is found on all leukocytes and is present in lymphomas, tumours of lymphoid tissue. Lymphomas can subsequently be sub-typed by immunocytochemistry for a variety of immunoglobulins

129

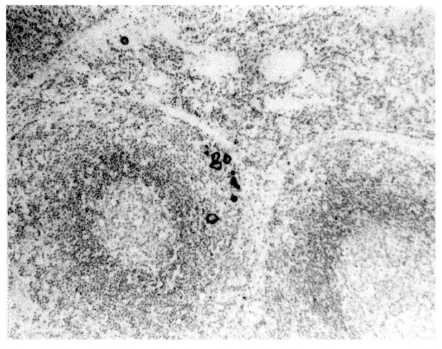

Figure 12.1: Lymph node, formalin-fixed, paraffin embedded. The section was pretreated with trypsin, then immunostained for cytokeratin with a peroxidase-labelled steptavidin method, the peroxidase developed with H_2O_2 and DAB, followed by counterstaining with haematoxylin. The section shows sparse metastic carcinoma cells. These would be difficult to find in a conventionally stained section.

and lymphocyte markers, many of which are only recently detectable in paraffin sections thanks to heat-mediated antigen retrieval. Frozen sections are now necessary for very few antigens commonly localized for diagnostic purposes.

Frozen sections are usually used for the diagnosis by immunofluorescence of kidney and skin auto-immune disorders through the pattern of immunoglobulin deposition in glomerular and skin basement membranes (see *Plate 1(a)*, p. 33) (Evans, 1986; Chu, 1986). Circulating auto-antibodies can also be identified by staining normal tissue with the patient's serum as primary antibody (Scherbaum *et al.*, 1986a,b).

Immunocytochemistry is labour-intensive and expensive, so tests should be used sparingly, and always with a mind to the clinical features of the case and after examination of conventionally stained material.

Panels of antibodies are generally used to reach or confirm a diagnosis. For further reading, see the work of Bosman (1990) and Leong (1993).

12.1.1 Controls

A laboratory stock should be kept of positive samples for each primary antibody used. If the same primary antibody is being used on several test samples in the same run, only one positive control preparation is necessary, but a negative control is needed for each test sample and preparative method. The long-term use of the same positive control samples will help to show up any deterioration of antibodies or other reagents.

Some laboratories use multi-tissue blocks for positive controls. In our opinion, this system is very useful for testing antibodies (see Section 7.1). However, when it is difficult to obtain normal and pathological blocks for controls, it is wasteful of tissue to immunostain it with irrelevant antibodies just because it is included in the multi-tissue control block; for example, a carcinoma with antibody to melanoma-specific antigen.

12.1.2   Choice of antibody

Among so many suppliers of antibodies supposed to label the same antigens, how is one to choose a suitable one? First of all, consult your colleagues, the literature and the suppliers' catalogues and if possible select an antibody that has already been used successfully for immunocytochemistry on formalin-fixed paraffin tissues (or whatever system you require it for). If no data are available, there is no alternative to testing it.

Cost may be a consideration, but it is often difficult to decide between the merits of a 'ready-to-use' diluted antibody and the concentrated immunoglobulin. The latter allows more flexibility in concentrations for different methods.

Reference points for choosing antibodies include Linscott's Directory (www.linscottsdirectory.com). This is a useful publication, also available on disc and continuously updated, which lists thousands of antibodies by species and clone with their suppliers' names and addresses. No further information is given – you will have to check further with the supplier.

A good free website is Immunoquery, http://immunoquery.com. This compiles published information and statistical data on immunodiagnosis of tumours and may help in deciding whether an antibody would be suitable. The website of the Society for Applied Immunohistochemistry (www.appliedimmuno.org) is another from which useful information can be gleaned. Many articles and books publish lists of antibody panels for diagnosis, although these can become outdated as new tests come into use.

12.1.3   Tips for diagnostic laboratories

This section lists technical tips for solving some of the problems that may arise in a diagnostic laboratory.

Shortage of material

1. *When only one section or one slide with several sections is provided for several immunostains:*

(a)   After bringing the section to water, blocking endogenous peroxidase and pre-treating with heat or protease as required, divide the section or isolate the individual sections with a water-resistant pen so that a different immunoreaction can be done on each segment. Remember to include an area for a negative control.

(b)   If you are confident in the method, carry out multiple immunostaining.

2.   *When an immunostaining run has been done and further tests are required, but there is no more tissue.* Provided that the previous run contained some negatively staining sections, which could be either negative controls or negative results, remove the coverslips, bring to water, remove the haematoxylin nuclear counterstain and re-stain with the required antibodies. If the correct pre-treatment has already been applied, you will not need to repeat it.

3.   *When only haematoxylin and eosin or other stained preparations are available.* Remove the coverslip, bring to water, de-stain and then re-stain with the immunocytochemical test. If antigen retrieval is required, it may be advisable to do this overnight at low-temperature if the preparation is not on suitably coated slides (see Section 3.8.3). If the tinctorial stain is required again, immunostain with a label that can be removed after examination and photography, e.g. peroxidase developed with an AEC method or alkaline phosphatase, so that the end-product can be dissolved in alcohol. The preparation can then be re-stained as before.

   This method is applicable to alcohol-fixed Papanicolaou-stained cytological preparations, but not, in our experience, to air-dried preparations that have been Giemsa-stained, even if they are fixed before the immunostain.

4.   *When cytospin preparations are scarce.* Try developing with a soluble end-product as above. Record the result then remove the stain and elute the immunoreagents with sodium dithionite (see Section 9.3.1) and re-stain. You should test for complete removal of antibodies before applying the new primary so this would be a prolonged experiment, only to be attempted if there is no easier way of getting new material. Re-use of a negative preparation would be simpler.

5.   The 'peel and stick' method is an alternative (Mehta and Battifora, 1993; Miller, 2001).

Failed reactions. If the positive control is satisfactory, but the test tissue is negative, it should be assumed that the negative result is correct. If both the positive control and the test tissue are negative, it should be assumed that something has gone wrong.

Check all steps. It may be possible to save the reaction (Section 8.1.2).

1.   If a layer has been omitted or the wrong anti-species reagent has been applied, it is usually possible to go back and repeat the reaction from the erroneous layer.

2. If the enzyme development has failed (e.g. through the blocking concentration of hydrogen peroxide being used instead of the much lower development concentration), it may be possible to return to the final layer and build up antibodies to reach another peroxidase-labelled reagent, which can be redeveloped.

Urgent requests. When time is short, try reducing the normal incubation period and increasing the reagent concentrations. The final result may not be optimal, but could give an answer. Many companies provide 'quick' or extra-strong development reagents which are useful for this strategy and it is a good idea to have some in stock if you can afford it. It will usually be easier and quicker to stain an urgent preparation 'by hand' rather than on a machine. It is advisable to stain a parallel preparation by the usual method in case the rapid method is unsatisfactory.

12.2 Research

The applications of immunocytochemistry to all kinds of research problem are too numerous to list here. Suffice it to say that the method can be combined with many other techniques, such as autoradiography, *in situ* hybridization of nucleic acids, histochemical staining and biogenic amine identification by formaldehyde-induced fluorescence. The technique is not confined to human tissue. Given the correct antibodies, antigens can be identified in animal specimens, both vertebrate and invertebrate, and also in plants. A point to remember is that an antibody raised to an antigen (say antigen x) of one species may be immunoreactive in a different species, but until the reacting antigen is proved to be identical with the original, the immunoreactivity must be referred to as 'x-like immunoreactivity', implying that the antigens are structurally related but may not be identical in every feature. It is particularly important to incorporate the standard controls in any experimental work. A further consideration is that the function of the localized antigens may be completely different in different species, depending as much on the way the receptors or targets have evolved as on the structure of the antigen.

12.3 Quantification

It is still very difficult, even impossible, to discover the absolute amount of antigen in a sample by reference to the intensity of the immunostain because of the variable conditions inherent in the method, but samples treated in the same way and immunostained simultaneously under the same conditions can be compared from the aspect of intensity of immunostain and/or area occupied by stained structures. Immunocytochemical reactions resulting in well-defined localization of antigens, with the reaction product contrasted with the background tissue, provide excellent material for comparative quantification of antigens. Both enzyme and fluorescence reactions are suitable. Particular attention must be given to the standardization of methods, adequate sampling and care in the selection of parameters to be measured. Computerized image analysis is essential for providing accurate and unbiased records. For further information see the work of Read and Rhodes (1993) and McBride (1995).

At the electron microscope level, immunolabelling with colloidal gold provides a better hope of equating the intensity of the labelling with the amount of antigen present because depth of penetration is discounted, the label being on the surface only. The number of particles can be compared between samples. Approaching an absolute quantification, the number of particles on a labelled sample can be counted and compared with the count from the same method applied simultaneously to an artifical sample containing a known concentration of the antigen against which the antibody has been calibrated. This type of comparison has also been attempted for light microscopical immuno-staining (Millar and Williams, 1982). However, available antigen in a fixed biological sample may be less than the true amount in the tissue.

12.3.1   Confocal microscopy

The confocal laser scanning microscope provides an easier way of measuring the total (immuno)fluorescence of a relatively thick specimen than physically sectioning and staining it and recording the results on each section. The microscope uses a laser beam of defined wavelength which is focused on a single point in the sample. The advantage of this is that the emitted fluorescent light from that point is focused to pass through a hole in a diaphragm before reaching the detecting apparatus. Any extraneous light from the surrounding tissue is blocked by the diaphragm and is not recorded. The sample is scanned at the first level to produce an image of a chosen area with much better definition than is usual in immunofluorescence preparations seen by conventional UV microscopy. The microscope is then refocused at the next level in the specimen and the process is repeated at suitable increments. A series of very highly resolved images through the specimen is thus produced and captured by the recording computer. These images may be viewed separately or combined to give a composite image of the total area of the immunostained structures in the specimen (*Plate 16*, p. 40). For instance, the area of nerves immunostained for a particular neuropeptide (such as neuropeptide Y) could be compared with the total nerve area immunostained for a general nerve marker (such as protein gene product 5.5) either in serial sections or by double immunofluorescence in the same specimen, and the same parameters compared in normal and pathological tissue. As an example, confocal microscopy was used to investigate the pattern of nerve growth in skin grafts (Gu *et al.*, 1995). The computer images may be artificially coloured as appropriate to the fluorophore if desired. For further reading on confocal microscopy, see the work of Matsumoto and Kramer (1994), Boyde (1995) and Sheppard and Shotton (1997).

12.3.2   Flow cytometry and fluorescent antibody cell sorting (FACS)

This technique allows populations of single cells to be screened by immunofluorescence for different cell surface membrane antigens. The cells can also be counted and (if required) separated into different populations on the basis of their ability to scatter light (dependent on the size of the cell) and on their fluorescence characteristics. The intensity of fluorescence can also be measured by a microfluorimeter.

Cells in suspension are allowed to pass singly and rapidly through the focus of a laser beam of the required wavelength or even, for multiple-stained cells, several beams of different wavelengths, so that several types of fluorescence can be assessed at the same time. Living cells can be labelled and, if a cell sorter is incorporated in the apparatus, they can be deflected and collected as different populations according to their fluorescence. It is also possible to disaggregate solid tissue and even to use tissue from fixed and embedded paraffin blocks, for example, to determine its DNA content (Camplejohn, 1992). The FACS machine can produce a graphic representation of the cell populations as a scatter-gram or histogram. The equipment required is large and expensive, and best used by a specialist in the technique. For further information see the work by Radbruch (1992) and Ormerod (1999).

12.3.3   Simpler methods of quantification

Computerized image analysers, confocal microscopes and FACS machines are complicated and costly. Once skill in their use has been acquired, they can be made to perform quantitative analyses with speed and accuracy, provided that the preparations they are asked to analyse have been carefully made and the correct measurements and statistics are applied. However, long-established ways of quantification, for instance by counting stained structures against a grid in the required number of microscope fields and other stereological techniques (Browne *et al.*, 1995), are still valid, though they may be less exciting than using elaborate equipment.

Supra-optimal dilution.   One simple way of comparing the antigen content of two samples that is available to any immunocytochemist is to use the supra-optimal dilution method devised by Vacca-Galloway (1985). The antibody is first titrated in a series of dilutions against a normal positive control to find the optimal dilution giving strong staining and low background and the supra-optimal dilution, at which staining is weak but still reliably demonstrates all the positive structures. To compare two antigen-containing structures, the antibody is used at supra-optimal and optimal dilutions sequentially on the same sections. At supra-optimal dilution, a structure with a large or normal amount of antigen will still be detected, although more weakly than usual because the antibody will not be in excess of the antigen so some antigen molecules will remain unstained. Where the antigen content is much lower than normal, there is less chance for the sparse antibody molecules to find their target and staining will be very much weaker or absent. At optimal dilution, both large and small amounts of antigen will be detected because the antibody will be in excess of the normal amount of antigen, ensuring that every available molecule of antigen is bound. Thus, by comparing intensity of staining or number of structures stained with the two dilutions, a comparison of antigen content in two samples can be made. Note that for this method to be successful, a relatively insensitive method such as the indirect or PAP method should be used rather than an ABC method which is so sensitive that it might be difficult to distinguish between high and low amounts of antigen (Sternberger and Sternberger, 1986). For an example of the practical use of this method, see the work by Roncalli *et al.* (1993).

Another use for this type of method is to examine the titre of different antibodies to the same antigen by comparing the end-point dilutions at which staining disappears. Serial sections containing the antigen would be necessary for this, or dot-blots containing a defined amount of antigen.

12.4   Non-immunocytochemical uses of labelled probes

Immunocytochemistry will certainly remain an invaluable tool for localizing tissue sites of the increasing number of identified antigens. No doubt, tissue preparation methods and antibody specificity will continue to be improved.

However, certain other, non-immunological methods of specific marking are now used. They use variously labelled probes to identify tissue constituents and include receptor localization, lectin histochemistry and *in situ* hybridization of nucleic acids. These methods are discussed briefly below because they share with immunocytochemistry many of the conditions and problems such as target preservation and method specificity. All the methods can be used in combination with immunocytochemistry to give additional information about the phenotype of the cells or tissues.

12.4.1   Receptor localization

The localization of radiolabelled hormones to their receptor site by autoradiography is well established (see, for example, Kuhar and Uhl, 1979; Wharton and Polak, 1993). With regard to receptor studies, immunocytochemical methods were sometimes used, even on fixed paraffin sections (Taylor *et al.* 1981), but there has been some doubt as to whether fixation and embedding can preserve all receptor sites adequately (Salih *et al.*, 1979), and so the use of fresh tissue has predominated (Goldsmith *et al.*, 1979; Willingham *et al.*, 1980; Buckley and Burnstock, 1986). Immunocytochemical studies have been made with anti-idiotypic antibodies, where the antibody mimics the ligand and binds to the receptor, but receptor proteins themselves are now being isolated in increasing numbers and used as antigens. An immunocytochemical method can then identify the receptor in the same way as any other antigen, particularly with the advent of heat-mediated antigen retrieval methods. Immunocytochemistry for oestrogen and its receptors was formerly possible only on frozen sections with antibodies designed for radioimmunoassay and therefore very expensive with the immunocytochemical technique because of the high concentration needed. With antigen retrieval, paraffin sections can be used and, with new antibodies, immunocytochemistry for oestrogen and progesterone receptors is now an important diagnostic test for hormone receptor status in breast cancer (see *Plate 2(a)*, p. 33). For discussion of many aspects of receptor localization, both immunocytochemical and autoradiographic, see Wharton (1996).

12.4.2   Lectin histochemistry

This is another non-antibody method of marking tissue components with a labelled probe, in this case a lectin. Lectins are plant or animal proteins that can attach to tissue carbohydrates (e.g. in glycoproteins) with a high degree of specificity according to the lectin and the carbohydrate group (Brooks *et al.*, 1996). Since the carbohydrates may be characteristic of a particular tissue, lectin binding may have diagnostic significance (Damjanov, 1987). A lectin from the gorseplant, *Ulex europaeus*, for example, is specific for α-L-fucose and a marker of human vascular endothelium. Tissue-bound lectins may be identified with antibodies to the lectin or, if they themselves are labelled, by any other suitable immunocytochemical technique, or by the label alone. The standard immunocytochemical labels are used. Lectins do not normally require pre-treatment of the fixed section, but for some carbohydrate groups, neuraminidase digestion may be helpful to remove sialic acid residues that may mask the reactive groups.

12.4.3 *In situ* hybridization of nucleic acids

As opposed to immunocytochemistry, which can only identify the site of cellular storage of an antigen, hybridization of the antigen's messenger ribonucleic acid (mRNA) with a specific labelled probe can identify its level of production within the cell. Thus, even if immunocytochemistry is unsuccessful because the antigen is being exported from the cell so rapidly that none remains to be localized, *in situ* hybridization can show a high level of mRNA. The probe is a synthesized complementary nucleic acid sequence incorporating an isotope (^{35}S, ^{32}P, ^{3}H) subsequently developed by autoradiography; ^{32}P-labelled riboprobes are the most successful. A non-isotopic label might be biotin, digoxygenin or fluorescein that is then amplified and revealed by a labelled antibody method. Probes are double-stranded DNA, single-stranded complementary RNA, synthetic oligonucleotides or peptide nucleic acids. For DNA probes, both the probe and nucleic acids in the sample must be dissociated into single strands and re-annealed so that the labelled probe is combined with its complementary strand. Cytoplasmic RNA may be hybridized without prior dissociation. The presence of a specific RNA can be confirmed by Northern blotting of a tissue extract. The conditions for *in situ* hybridization are more stringent than for immunocytochemistry. For details, see the work of Leitch *et al.* (1994).

In situ polymerase chain reaction. The polymerase chain reaction (PCR) is a means of amplifying a small amount of DNA or RNA in a tissue to the level at which it can be identified with a labelled probe. It is usually done on extracts of tissue samples run on a gel, but can be done on frozen or paraffin sections (*in situ* PCR), the amplified nucleic acids being localized in the same way as for classical *in situ* hybridization.

References

Bosman FT. (1990) General approach to tumour markers in diagnostic pathology. In *Histochemistry in Pathology* (eds IM Filipe, B Lake). Churchill Livingstone, Edinburgh, pp. 49–60.

Boyde A. (1995) Confocal optical microscopy. In *Image Analysis in Histology, Conventional and Confocal Microscopy* (eds R Wootton, DR Springall, JM Polak). Cambridge University Press, Cambridge, pp. 151–196.

Brooks SA, Leathem AJC, Schumacher U. (1996) *Lectin Histochemistry*, Microscopy Handbook 36. BIOS Scientific Publishers, Oxford.

Browne MA, Howard CV, Jolleys GD. (1995) Principles of stereology. In *Image Analysis in Histology, Conventional and Confocal Microscopy* (eds R Wootton, DR Springall, JM Polak). Cambridge University Press, Cambridge, pp. 96–120.

Buckley NJ, Burnstock G. (1986) Localisation of muscarinic receptors on cultured myenteric neurons: a combined autoradiographic and immunocytochemical approach. *J. Neurosci.* **6**, 531–540.

Camplejohn RS. (1992) Flow cytometry in clinical pathology. In *Diagnostic Molecular Pathology, A Practical Approach,* Vol. I. (eds CS Herrington, J O'D McGee). IRL Press, Oxford, pp. 239–257.

Chu AC. (1986) Immunocytochemistry in dermatology. In *Immunocytochemistry, Modern Methods and Applications* (eds JM Polak, S Van Noorden). Butterworth–Heinemann, Oxford, pp. 618–637.

Damjanov I. (1987) Biology of disease, lectin cytochemistry and histochemistry. *Lab. Invest.* **57**, 5–20.

Evans DJ. (1986) Immunohistology in the diagnosis of renal disease. In *Immunocytochemistry, Modern Methods and Applications* (eds JM Polak, S Van Noorden). Butterworth–Heinemann, Oxford, pp.638–649.

Goldsmith PC, Cronin MJ, Weiner RI. (1979) Dopamine receptor sites in the anterior pituitary. *J. Histochem. Cytochem.* **27**, 1205–1207.

Gu XH, Terenghi G, Kangesu T, Navsaria HA, Springall DR, Leigh IM, Green CJ, Polak JM. (1995) Regeneration pattern of blood vessels and nerves in cultured keratinocyte grafts assessed by confocal laser scanning microscopy. *Br. J. Dermatol.* **132**, 376–383.

Kuhar MT. Uhl GR. (1979) Histochemical localization of opiate receptors and the enkephalins. In *Neurochemical Mechanisms of Opiates and Endorphins* (eds HH Loh, DH Ross). Raven Press, New York, pp. 53–68.

Leong AS-Y. (1993) *Applied Immunohistochemistry for the Surgical Pathologist.* Edward Arnold, Melbourne.

Leitch AR, Schwarzacher T, Jackson D, Leitch IJ. (1994) *In Situ Hybridization,* Microscopy Handbook 27. BIOS Scientific Publishers, Oxford.

McBride JT. (1995) Quantitative immunocytochemistry. In *Image Analysis in Histology, Conventional and Confocal Microscopy* (eds R Wootton, DR Springall, JM Polak). Cambridge University Press, Cambridge, pp. 339–354.

Matsumoto B, Kramer T. (1994) Theory and applications of confocal microscopy. *Cell Vision* **1**, 190–198.

Mehta P, Battifora H. (1993) How to do multiple immunostains when only one tissue slide is available. The 'Peel and Stick method'. *Appl. Immunohistochem.* **1**, 297–298.

Millar DA, Williams ED. (1982) A step-wedge standard for the quantification of immunoperoxidase techniques. *Histochem. J.* **14**, 609–620.

Miller RT. (2001) Technical Immunohistochemistry. www.appliedimmuno.org

Ormerod MG. (ed.) (1999) *Flow Cytometry,* 2nd edition, Microscopy Handbook 44. BIOS Scientific Publishers, Oxford.

Radbruch A. (ed.) (1992) *Flow Cytometry and Cell Sorting.* Springer-Verlag, Heidelberg.

Read NG, Rhodes P. (1993) Techniques for image analysis. In *Immunocytochemistry, A Practical Approach* (ed. JE Beesley). Oxford University Press, Oxford, pp. 127–149.

Roncalli M, Springall DR, Maggioni M, Moradoghli-Haftvani A, Winter RJ, Zhao L, Coggi G, Polak JM. (1993) Early changes in the calcitonin gene-related peptide (CGRP) content of pulmonary endocrine cells concomitant with vascular remodeling in the hypoxic rat. *Am. J. Respir. Cell Mol. Biol.* **9**, 467–474.

Salih H, Murthy GS, Friesen HG. (1979) Stability of hormone recptors with fixation: implications for immunocytochemical localization of receptors. *Endocrinology* **105**, 21–26.

Scherbaum WA, Mirakian R, Pujol-Borrell R, Dean BM, Bottazzo GF. (1986a) Immunocytochemistry in the study and diagnosis of organ-specific auto-immune diseases. In *Immunocytochemistry, Modern Methods and Applications* (eds JM Polak, S Van Noorden). Butterworth–Heinemann, Oxford, pp. 456–476.

Scherbaum WA, Blaschek M, Berg PA, Doniach D, Bottazzo GF. (1986b) Spectrum and profiles of non-organ-specific auto-antibodies in auto-immune disease. In *Immunocytochemistry, Modern Methods and Applications* (eds JM Polak, S Van Noorden). Butterworth–Heinemann, Oxford, pp. 477–491.

Sheppard C, Shotton D. (1997) *Confocal Laser Scanning Microscopy*, Microscopy Handbook 38. BIOS Scientific Publishers, Oxford.

Sternberger NH, Sternberger LA. (1986) The unlabeled antibody method. Comparison of sensitivity of peroxidase–antiperoxidase with avidin–biotin complex method by a new mode of quantitative immnocytochemistry. *J. Histochem. Cytochem.* **34**, 599–605

Taylor CR, Cooper CL, Kurman RJ, Goebelsman U, Markland FS. (1981) Detection of estrogen receptor in breast and endometrial carcinoma by the immunocytochemical technique. *Cancer* **47**, 2634–2640.

Vacca-Galloway LL. (1985) Differential immunostaining for substance P in Huntington's diseased and normal spinal cord: significance of serial (optimal, supra-optimal and end-point) dilutions of primary antiserum in comparing biological specimens. *Histochemistry* **83**, 561–569.

Wharton J. (ed) (1996) Receptor localization and analysis. *Histochem. J.*, **28**, 727–823.

Wharton J, Polak JM. (eds) (1993) *Receptor Autoradiography, Principles and Practice.* Oxford University Press, Oxford.

Willingham MC, Maxfield FR, Pastan I. (1980) Receptor-mediated endocytosis of α_2-macroglobulin in cultured fibroblasts. *J. Histochem. Cytochem.* **28**, 818–823.

Appendix: Technical Notes

The methods described here are in use in the authors' laboratories. Many satisfactory variations are to be found in the literature, and there are numerous reliable suppliers of reagents in addition to those quoted. Where water is mentioned it is assumed to be distilled or deionized unless tap water is specified.

A.1 Buffers for diluting antibodies and rinsing

A.1.1 Phosphate-buffered normal saline (PBS), 0.01 M, pH 7.2–7.4

Reagents
Sodium chloride, 8.7 g
Potassium dihydrogen phosphate, 0.272 g
Disodium hydrogen phosphate, 1.136 g
   or
Disodium hydrogen phosphate.2 H_2O, 1.41 g
   or
Disodium hydrogen phosphate.12 H_2O, 2.83 g

Method
Dissolve salts separately in water then mix, make up to 1 l and check pH. A 10-times concentrated stock solution (0.1 M) may be made and then diluted with water for use. The pH of the diluted solution should be checked as it will differ from that of the concentrate.

A.1.2 Tris-buffered normal saline (TBS), 0.05 M, pH 7.6

Reagents
Tris(hydroxymethyl)methylamine, 6.07 g

Sodium chloride, 8.7 g
Concentrated hydrochloric acid

Method
Dissolve the Tris and sodium chloride in 900 ml water. Add concentrated HCl until the pH reaches 7.6. Make up to 1 l with water.

A.2 Antibody diluent and storage of antibodies

PBS (see Section A.1.1) or TBS (see Section A.1.2) containing 0.1% BSA (Sigma A 4503) and 0.1% sodium azide.

Note: Antibodies may be diluted in simple PBS or TBS, but for highly diluted antibodies, add 0.1% of (approximately globulin-free) BSA or a suitable normal serum to provide a high concentration of non-antibody protein which can compete with the antibody for non-specific binding sites on the walls of the storage vessel, and also in the tissue. Monoclonal antibodies in medium containing fetal calf serum may be frozen in this medium.

For long-term storage at working dilution at 4°C, add 0.1% sodium azide as a preservative.

Note: Do *not* add sodium azide to enzyme-labelled reagents as it inhibits enzyme activity. To store peroxidase-linked reagents at working dilution at 4°C, use Protexidase (ICN 980631) as a diluent.

Most antibodies may be stored frozen at –20°C, but this should be tested before the entire stock is frozen. The antibodies should be divided into suitable aliquots and snap-frozen in liquid nitrogen before being put in the freezer. Avoid repeated thawing and freezing, and do not freeze antibodies at dilutions higher than 1/100. They should be diluted in buffer with protecting protein.

Antibodies diluted 1:1 with glycerine may be stored at –20°C without freezing. The dilution must be remembered when the working solution is prepared.

A.2.1 Double dilutions

1. Calculate the volume of each dilution needed to cover the preparation – say 100 µl. Allow a little extra for safety.
2. Label 8 vials with the dilutions required. In the first vial, make double the volume of the most concentrated sample, say 300 µl of 1/50 dilution (6 µl of antiserum added to 300 µl of diluent. As this is an arbitrary quantity it is not necessary to add 6 µl to 294 µl of diluent). Mix gently, but thoroughly.
3. Dispense 150 µl of diluent into each of the remaining vials.

4. Take 150 μl from the 1/50 dilution and add it to the 150 μl of diluent in the second (1/100) vial. Mix as before, then take 150 μl and add to the 150 μl of diluent in the next (1/200) vial and so on until the final dilution is reached. You will have excess of the highest dilution, but this does not matter.

Note: It is much easier to dilute a series in this way, ensuring uniform dilution from an initial solution than to add decreasing quanties of precious antibody from the stock to the small volume required for immunostaining a section.

A.3   Adherence of preparations to slides

A.3.1   Coating slides with *poly*-L-lysine (Huang *et al.*, 1983)

Reagents
Poly-L-lysine hydrobromide, molecular weight 150 000–300 000 (Sigma P1399).

Method
1. Dissolve in water to make 0.1% w/v solution. Aliquot into 0.5–1 ml portions and store frozen at –20°C.
2. For use, thaw an aliquot completely and stir before use. Replace unused solution in the freezer for storage.
3. Apply a small drop (5–10 μl) to one end of a clean slide, placed flat on the bench. Rock the end of another slide in the drop to distribute the solution along the edge. Push the second slide along the first slide at an angle of about 45°, applying a little pressure to spread the *poly*-L-lysine as a thin, even film over the whole surface. Interference colours should be seen as it is spread. If the solution spreads in droplets instead of an even film, the slide is not clean enough. Slides purchased as pre-cleaned are usually satisfactory but if cleaning is required, soak the slides in acid-alcohol overnight then rinse well in alcohol. The film will dry rapidly and the slide is ready for use. Label the coated side, as the dry film is invisible. Slides can be coated in large batches and stored indefinitely at room temperature.

Alternative method
A solution of *poly*-L-lysine, 0.1% aq. w/v (Sigma P 8920), is a little more expensive, but contains a preservative and can be stored at room temperature.
    For cell preparations a lower molecular weight polymer (30 000) can

be used, if preferred. Dip the slides in the solution and allow them to dry before applying drops of suspended cells which will then settle on the slides. *Cytospin preparations* can be made on either type of slide.

A.3.2   Coating slides with silane

Reagents
3-Aminopropyltriethoxy silane (APES) (Sigma, A 3648) and acetone.

Method
1. Place slides in a rack and immerse in acetone for 5 min.
2. Immerse in a solution of 2% silane in acetone v/v for 5 min.
3. Rinse in two consecutive baths of acetone for 5 min each. Allow to dry and store at room temperature indefinitely.

A.4   Blocking endogenous peroxidase reaction

A.4.1   Paraffin sections

Hydrogen peroxide, standard method

Reagents
Stock solution: 30% aq. (w/v) hydrogen peroxide. This must be stored at 4°C and renewed if it deteriorates (as shown by failure of the DAB reaction). If solutions of lower concentration only are available, adjust the quantity required accordingly. Working solutions must be freshly prepared and discarded after use.

Method
1. De-wax and bring to water through graded alcohols.
2. Immerse the sections in 0.6% hydrogen peroxide (2 ml 100 ml^{-1} water) for 15 min (a shorter time may be adequate). Buffer or methanol may be used instead of water, but both are more expensive. Methanol is itself a blocker of peroxidase and may be used with up to 3% hydrogen peroxide for persistent enzyme activity.
3. Rinse in water, then buffer and proceed with the immunostaining method. Pre-treatment of the preparations with enzymes or heat may be done before or after the peroxide block.

For sections containing quantities of red blood cells
(Heyderman, 1979)
Acid haematein is bleached with strong hydrogen peroxide. Peroxidase is

further inhibited with periodic acid, and any aldehyde groups created by periodic acid oxidation or left in the tissue by the fixative are reduced to non-sticky alcohol groups with borohydride. Take care – this method may damage the immunoreactivity of some antigens (e.g. LCA). Test before use.

Reagents
(a)   30% aq. hydrogen peroxide: 6 ml in 100 ml water (i.e. 1.8% final concentration)
(b)   2.5% aq. periodic acid
(c)   0.02% aq. potassium (or sodium) borohydride (this solution must be freshly made)

Method
1.   Immerse slides in hydrogen peroxide solution for 5 min.
2.   Rinse in tap water.
3.   Immerse in periodic acid solution for 5 min.
4.   Rinse in tap water.
5.   Immerse in borohydride solution for 2 min.
6.   Rinse in tap water and proceed with immunostaining method.

A.4.2   Milder methods for cryostat sections and whole-cell preparations

Hydrogen peroxide in methanol/PBS

Reagents
(a)   30% hydrogen peroxide
(b)   70% methanol in PBS

Method
Immerse preparations in 0.3% hydrogen peroxide in 70% methanol in PBS (1 ml 30% H_2O_2 per 100 ml methanol/PBS) for 30 min.

Azide and nascent hydrogen peroxide (Andrew and Jasani, 1987)

Reagents
(a)   PBS (pre-warmed at 37°C), 100 ml
(b)   1 mM sodium azide (0.65%), 1 ml
(c)   10 mM glucose (18% aq.), 1 ml
(d)   Glucose oxidase (Sigma G 6891) to give 1 U ml^{-1} (quantity depends on batch volume – around 80–100 μl)

Method
1.   Just before use, mix all reagents and immerse preparations for 1 h at 37°C.
2.   Rinse in water, then buffer, and proceed with immunostaining method.

A.4.3   **Blocking endogenous biotin** (Wood and Warnke, 1981)

Standard reagents
(a)   Avidin/streptavidin, 1 mg ml^{-1} in buffer
(b)   Biotin, 0.1 mg ml^{-1} in buffer

Alternative reagents (adapted from Miller and Kubier, 1997, Miller *et al.*, 1999)
(a)   Avidin. One large egg white mixed well with 100 ml of water. Centrifuge to remove solids. Add 0.1% sodium azide and store at 4°C in 20 ml aliquots (Universal bottles).
(b)   Biotin. Skimmed milk or 5% powdered, fat-free, dried milk in PBS with 0.05% Tween 20. Add 0.1% sodium azide and store in 20 ml aliquots at −20°C. Keep current use supply at 4°C.

Method
1.   Incubate the preparation with avidin (a) for 20 min.
2.   Wash in buffer.
3.   Incubate with biotin (b) for 20 min.
4.   Rinse in buffer and continue with background blocking for the immunostaining method.

A.5   Enzyme pre-treatment

Note: Batches of enzymes vary in their potency, so a new batch may have to be evaluated to establish optimal digestion times. Different antigens require different digestion times, and the length of time in fixative (formalin) can also affect the time needed for antigen retrieval. Proteolytic digestion can be done before or after blocking endogenous peroxidase.

A.5.1   Trypsin

Reagents
(a)   Trypsin (crude porcine), approximately 435 USP units mg^{-1} (ICN 150213)
(b)   Calcium chloride, 0.1% in water or 0.005 M Tris/HCl buffer or Tris/HCl-buffered 0.9% sodium chloride (TBS), pH 7.6–8.00. This solution is conveniently kept in bulk at 37°C so that it is already warm for use.
(c)   0.1 M (0.4%) sodium hydroxide (aq.)

Method
1. Dissolve trypsin in the calcium chloride to make a 0.1% solution. The trypsin will not dissolve completely until the pH nears 7.8.
2. As quickly as possible (to avoid cooling) adjust the pH to 7.8 with 0.1 M sodium hydroxide.
3. Immerse the slides in the trypsin solution and place at 37°C for the pre-determined optimal time (usually between 10 and 20 min).
4. Rinse the slides well in running tap water.
5. Proceed with the immunostaining method.

Note: An alternative method is to make a much smaller volume of solution and apply it as drops on the preparations, but immersing the slides in the solution provides an even and uniform digestion, which is easier to standardize. Trypsin is available in pre-weighed tablet form from many companies.

A.5.2   Protease

Reagents
(a)   Protease XXIV, bacterial (EC 3.4.21.62), 7–14 U mg^{-1} (Sigma P 8038)
(b)   PBS, pH 7.2–7.4 (see Section A.1), stock solution pre-warmed at 37°C.

Note: Because of the cost of this enzyme, it is applied to the preparations in small drops.

Method
1. Make a 0.05–0.5% solution of protease in PBS (e.g. 1–10 mg in 2 ml). There is no need to adjust the pH.
2. Apply as drops on the preparations. Incubate at 37°C for the appropriate time.
3. Wash well in tap water and proceed with the immunostaining method.

A.5.3   Pepsin

Reagents
(a)   Pepsin, from porcine stomach (EC 3.4.23.1), 1:2500, 600–1000 U mg^{-1} (Sigma P 7125)
(b)   Hydrochloric acid, 0.01 M at 37°C

Method
1. Make a 0.4% solution of pepsin in HCl.
2. Incubate preparations for the appropriate time at 37°C.
3. Wash well in tap water and proceed with the imunostaining method.

Note: this enzyme is relatively cheap, so the immersion method could be used.

A.5.4   Neuraminidase

Reagents
(a)   Neuraminidase (Sialidase) Type V, from *Clostridium perfringens* (EC 3.2.1.18), 0.5–6 U mg^{-1} (Sigma N-2876)
(b)   0.1 M sodium acetate/acetic acid buffer, pH 5.5, stock solution stored at 37°C
(c)   0.1% aq. calcium chloride

Method
1.   Make a solution of neuraminidase of 1 U ml^{-1} in acetate buffer and store frozen in 100 μl aliquots each containing 0.1 U.
2.   For use, thaw an aliquot, add 800 μl of warm acetate buffer and 100 μl of calcium chloride solution (0.1 U neuraminidase ml^{-1}).
3.   Apply as drops on the preparations and incubate for the standard time (about 10 min) at 37°C.
4.   Wash well in tap water and proceed with the immunostaining method.

A.6   Heat-mediated antigen retrieval using a microwave oven

Equipment
(a)   Microwave oven with revolving plate, timer and choice of watt settings
(b)   Plastic container to take slide carrier(s), volume > 500 ml (e.g. sandwich box), with lid (optional)
(c)   Plastic slide carrier

Citrate buffer, pH 6.0

Reagents
For 2 l of 0.01 M citrate buffer, pH 6.0:

(a) 3.8 g citric acid
(b) 2 M (8% aq.) sodium hydroxide

Method
1.   Dissolve the citric acid in 1.9 l of water.
2.   Add the sodium hydroxide solution until the pH reaches 6 (20–25 ml). Make up to 2 l with water. The buffer can be kept at room temperature for some days.

Tris/EDTA, pH 9.0

Reagents
(a) Tris(hydroxymethyl)methylamine, 12 g
(b) Ethylene-diamine tetra-acetic acid, disodium salt, dihydrate 1 g
(c) 1 M hydrochloric acid 500 ml
(d) Water 500 ml

Method
1. Dissolve the salts in the water.
2. Add the hydrochloric acid
3. Check the pH

Note: This formula is 10 × concentrated. Keep the stock solution at 4°C and dilute 1:9 (v:v) for use.

Microwaving

Method
1. To standardize the method and ensure a supply of warm water to compensate for evaporation, heat 500 ml of buffer in the container to be used for the slides together with 200 ml of water in a plastic beaker for 2 min at 750 W. At the end of this time, remove the beaker of water.
2. Place the slides (de-waxed and brought to water) in a plastic slide carrier and immerse them completely in the warm buffer. The buffer solution must cover the slides completely. Cover the container loosely with its lid or with 'cling-film'.
3. Microwave at 750 W for 5 min (The solution must boil).
4. Check the level of the solution and make up to the original level if necessary with the warm water.
5. Repeat steps 5 and 6 for the required length of time (in our experience between 10 and 30 min).
6. Remove the container from the microwave oven, place it in a sink and run in cold tap water slowly until the solution is cool enough to remove the slide carrier without danger of buffer salts precipitating upon the sections as they dry.
7. Rinse the slides in water and continue with the immunostaining method.

Note 1: If sections requiring different heating times are included in the same run, begin with the ones that need the longest time and add the others at appropriate stages. This avoids having to remove slides from boiling solution, resulting in drying of the section.
Note 2: There are many ways of organizing heat-mediated antigen retrieval. The one cited is in use in our laboratories. A pressure cooker or autoclave may be used instead of a microwave oven (see main text for references).

Pressure cooking

Equipment
(a)   Stainless steel pressure cooker
(b)   Electric hot plate
or
(a)   Plastic pressure cooker
(b)   Microwave oven

Method
1.   Place enough buffer in the pressure cooker on hot plate or (plastic cooker only) in microwave oven and bring to boiling.
2.   Immerse the slides (plastic slide rack for microwave oven), close cooker and bring to pressure.
3.   Maintain pressure for 2 minutes, or longer if necessary, then remove cooker from heat source. Allow to cool by standing until pressure is reduced or cool rapidly under running water. Remove lid.
4.   When slides can be removed without danger of drying, rinse them in water and continue with the immunostaining method.

A.7   Enzyme development methods

Note: DAB and many of the other chemicals used in these methods are toxic or potentially carcinogenic. Wear gloves, use a fume cupboard when possible and take due care in disposing of waste. See Note 2 below.

A.7.1   Peroxidase

Diaminobenzidine standard method
The method given is for 'home-made' DAB with the immersion method. If you are using DAB tablets or one of the stable DAB solutions now available and applying the solution as drops on the preparations, follow the instructions supplied by the manufacturer. The method given below may be scaled up or down according to the volume required. (See Note 3 below for a safe and convenient method of preparing and storing aliquots of DAB.)

Reagents
(a)   3,3'-Diaminobenzidine tetrahydrochloride (DAB), 50 mg (e.g. Sigma D5637)
(b)   PBS or TBS, 100 ml
(c)   Hydrogen peroxide (30%), 30 μl (final concentration 0.01–0.03%)

Method

1. Add the DAB to the buffer and mix well. The solution should be clear or a light straw colour.
2. Just before use, add the H_2O_2.
3. Immerse the slides (previously in buffer) for 10 min.
4. Rinse the slides in buffer and examine microscopically. The end-product of reaction is dark brown. Shorter or longer incubation times may be necessary, but after about 15 min no further useful reaction is likely to occur as the reaction product masks the site of reaction. Longer incubation may result in background staining.
5. Rinse well in running tap water.
6. Counterstain lightly in haematoxylin (if too dark, contrast will be impaired). Differentiate in acid-alcohol (1% HCl in 70% alcohol), 'blue' in tap water or Scott's tap water substitute, dehydrate through graded alcohols, clear in solvent and mount in synthetic mountant.

Note 1: Re-use of DAB solution. The DAB/H_2O_2 solution may be re-used several times and will remain useful for at least an hour (better kept in the dark). When it starts to go brown (oxidizes), discard it.

Note 2: Disposal of DAB solution. For safe disposal of DAB solution, add a few drops of household bleach (containing sodium hypochlorite). The DAB is oxidized (the solution becomes black) and can be washed down the sink drain with plenty of water. It is important to wash the DAB container very well after this to remove all traces of bleach which would compromise the next DAB reaction resulting in the solution becoming brown (and useless) on addition of the H_2O_2.

Alternative method, approved by FDA, USA (Lunn and Sansome, 1990). Add about 15 ml of concentrated sulphuric acid to about 85 ml of water, slowly, with stirring. Then add about 4 g of potassium permanganate. Add to the used DAB solution and leave overnight. Neutralise with sodium hydroxide (care – the temperature will rise), then discard.

Note 3: Safe use of DAB. DAB is probably most dangerous in the powder form which could be inhaled, but is much cheaper in bulk than in tablets. To avoid continually weighing small quantities of DAB, buy pre-weighed amounts (e.g. 5 or 25 g), dissolve the entire amount in distilled water to give a solution containing 50 mg DAB ml^{-1} (stir covered for 15 min in a fume cupboard), then dispense into vials containing the appropriate volume for your use, for example 5 ml to make 500 ml or 1 ml to make 100 ml final solution. Cap the vials and store frozen (–20°C). For use, thaw a vial and wash the contents into the correct volume of buffer. This method also ensures uniformity of developing solution for each batch of DAB powder (Pelliniemi *et al.*, 1980).

Aminoethyl carbazole (van der Loos, 1999, adapted from Graham *et al.*, 1965)

Note: This method gives a red end-product that is alcohol-soluble. An aqueous mountant must be used.

Reagents
(a)   3-Amino-9-ethyl carbazole (AEC) (Sigma A 5754)
(b)   *N,N*-Dimethyl formamide (DMF).
      Stock solution: 1.0% AEC in DMF. (Store at room temperature protected from light)
(c)   0.05 M sodium acetate/acetic acid buffer, pH 5.2
(d)   30% hydrogen peroxide

Method
1.   Wash preparations briefly in water then equilibrate in acetate buffer (c) for a few minutes.
2.   Incubating solution: add 2.5 ml stock AEC solution to 47.5 ml acetate buffer (c). Mix well (the solution will be slightly turbid). Filter. Just before use, add 20 µl hydrogen peroxide (d).
3.   Incubate for 10 to 20 minutes. At intervals during this time, rinse in buffer and check microscopically. Re-apply incubating solution if necessary. Stop the reaction when a satisfactory red end-product of reaction has been reached.
4.   Rinse in acetate buffer then in tap water.
5.   Counterstain nuclei lightly with Mayer's haemalum (to avoid the need for differentiation in acid alcohol since the reaction product is alcohol-soluble). If another haematoxylin is used and differentiation is necessary, use aqueous 1% hydrochloric acid.
6.   Blue in tap water or Scott's tap water substitute.
7.   Mount in an aqueous mountant, e.g. Hydromount (National Diagnostics) or, for a permanent preparation, Aquaperm (ThermoShandon). Apply this mountant directly to the preparations in a thin layer. Dry (protected from dust) at room temperature or at 37°C, then cover with synthetic mountant and coverslip.

4-Chloro-1-naphthol (modified from Nakane, 1968)
This method gives a blue-grey reaction product, soluble in alcohol.

Reagents
(a)   4-Chloro-1-naphthol (Sigma C 8890), 30–40 mg
(b)   Alcohol, 0.2–0.5 ml
(c)   PBS or TBS, 100 ml
(d)   Hydrogen peroxide (30%), 50–100 µl

Method
1.   Dissolve 4-chloro-1-naphthol in alcohol.
2.   Add with stirring 100 ml of PBS or TBS containing 50–100 µl H_2O_2. A white precipitate is formed. Heat to 50°C, filter through coarse filter paper into a Coplin jar and use the filtrate while hot.
3.   Incubate preparations for 5–10 min, checking microscopically.
4.   Rinse in water, counterstain in Carmalum (red nuclei) and mount in an aqueous medium (see AEC method above).

Phenol tetrazolium method (Murray *et al.*, 1991)
Note: This method, as sensitive as the usual DAB method, produces a dark blue insoluble reaction product. The pseudo-peroxidase of red blood cells does not react (though other endogenous peroxidases do).

Reagents
(a) 0.05 M Tris/HCl buffer, pH 7.6
(b) Phenol, 1 mg ml^{-1} in the buffer
(c) Reduced nicotinamide adenine dinucleotide (NADH), 1mg ml^{-1} in the buffer
(d) Nitro blue tetrazolium (NBT), 1.5 mg ml^{-1} in the buffer
(e) Hydrogen peroxide, final concentration in the buffer 0.02% (0.66 μl of 30% H$_2$O$_2$ ml^{-1} of buffer)

Method
1. Incubate preparations at 37°C for 10 min.
2. Wash in water.
3. Counterstain with carmalum or methyl green.
4. Air-dry and mount in synthetic resin.

A.7.2   Alkaline phosphatase

To give a blue end-product (Burstone, 1961)

Reagents
(a) Naphthol AS-MX phosphate, sodium salt (Sigma N 5000), 20 mg
(b) *N,N*-Dimethyl formamide, 0.4 ml
(c) 0.1 M Tris/HCl buffer, pH 8.2, 19.6 ml
(d) Levamisole hydrochloride (inhibitor of endogenous alkaline phosphatase except intestinal isoenzyme) (Sigma L 9756)
(d) Fast Blue BB salt (Sigma F 3378)

Method
Stock solution:

1. Dissolve the naphthol AS-MX phosphate in the dimethylformamide (use a glass vessel, not a plastic one).
2. Add the Tris buffer quickly with stirring.
Note: This stock substrate solution may be stored at 4°C for several weeks. Alternatively, it may be made with 1 mg naphthol AS-MX phosphate ml^{-1} of buffer and stored frozen in 1 ml aliquots. For use, thaw 1 ml (1 mg) and add 4 ml of buffer to give 5 ml.
3. Add levamisole to give a 1 mM solution (approx. 1.2 mg per 5 ml).
Note: Levamisole may be added to the Tris buffer and stored as a stock solution before mixing with the aliquot of substrate. In this case, to compensate for the 1 ml of substrate that does not contain the inhibitor, use 1.45 mg levamisole ml^{-1} of stock buffer.

4. Just before use, add Fast Blue BB, 1 mg ml^{-1}. Mix well and filter on to the preparations.
5. Incubate at room temperature for 5–15 min checking microscopically for the bright blue reaction product. If the reaction is slow, put the preparations at 37°C.
6. Wash in running tap water, counterstain in carmalum if required and mount in an aqueous mountant (reaction product is alcohol-soluble).

To give a red end-product (Fast Red TR method) (Burstone, 1961)
As above, using Fast Red TR salt (Sigma F 8764) instead of Fast Blue BB, and a blue nuclear counterstain (Mayer's haemalum).

Indoxyl (BCIP) method to give a less soluble blue-brown product
(de Jong *et al.*, 1985)
Note: This is said to be the most sensitive of the alkaline phosphatase development methods. The end-product is less soluble than in the methods given above, but it is probably still advisable to use an aqueous mountant. The substrate is expensive.

Reagents
Solution A:
(a) 5-Bromo-4-chloro-3-indolyl phosphate, disodium salt (Sigma B 6149), 5 mg dissolved in
(b) *N,N*-Dimethyl formamide, 0.1 ml
Solution B:
(c) Tetra nitro blue tetrazolium (TNBT) (Sigma T 4000), 10 mg dissolved in 0.2 ml dimethyl formamide. Add
(d) 0.2 M Tris buffer, pH 9.5, 0.1 ml
Solution C:
(e) 0.2 M Tris buffer, pH 9.5, containing 10 mM magnesium chloride and 10 mM levamisole, 30 ml.

Method
1. Mix together solutions A,B and C.
2. Filter and incubate preparations for 30 min at room temperature.
3. Rinse in water, counterstain nuclei with Mayer's carmalum and mount with an aqueous mountant.

A.7.3 Glucose oxidase (Suffin *et al.*, 1979)

Reagents
(a) β-D-Glucose, final concentration 7.5 mg ml^{-1} (42 mM)
(b) Nitro blue tetrazolium (NBT) (Sigma N 6876), 0.5 mg ml^{-1} (0.7 mM)

(c)   Phenazine methosulphate (PMS) (Sigma P 9625), 0.1 mg ml^{-1} (0.32 mM)

(d)   0.1 M phosphate buffer, pH 6.9

Method

For 5 ml incubating solution:

1.   Pre-heat 37 mg glucose with 3 mg NBT in 5 ml phosphate buffer to 37°C.
2.   Weigh 10 mg PMS and dissolve in 1 ml of buffer. Add 50 µl of this to the glucose/NBT solution.
3.   Incubate preparations at 37°C for about 1 h. The reaction product is dark blue.
4.   Wash in buffered formalin for 5 min then rinse well in tap water. Counterstain nuclei red with carmalum and mount in aqueous medium (or counterstain with neutral red, dehydrate, clear and mount in synthetic medium).

A.7.4   *β*-D-**Galactosidase** (Bondi *et al.*, 1982)

Reagents

Solution A:

(a)   5-Bromo-4-chloro-indolyl-*β*-D-galactopyranoside (X-Gal) (BCIG) (Sigma B 4252), 50 mg dissolved in
(b)   *N,N*-Dimethyl formamide, 0.5 ml

Solution B:

(c)   50 mM potassium ferricyanide, 0.5 ml
(d)   50 mM potassium ferrocyanide, 0.5 ml
(e)   PBS, pH 7.0–7.5, containing 1 mM magnesium chloride, 7 ml

Method

1.   Add 0.05 ml solution A to 2.276 ml solution B.
2.   Incubate preparations for 1 h at 37°C. The reaction product is turquoise blue.
3.   Wash in tap water, counterstain nuclei with carmalum or neutral red. Dehydrate through graded alcohols, clear and mount in synthetic medium.

Note 1: The pH for this reaction is designed for the bacterial enzyme used to label the immunoreagent. The histochemical reaction for the mammalian enzyme is done at pH 4.0.

Note 2: Solutions A and B and the final incubating solution can be stored frozen.

A.8   Intensifying the peroxidase/DAB reaction product

A.8.1   Following standard development

Copper sulphate (Hanker *et al.*, 1979)

Reagents
(a)   Copper sulphate
(b)   0.85% sodium chloride

Method
1.   Dissolve copper sulphate in sodium chloride to 0.5%.
2.   Immerse the preparations for 2–10 min, checking microscopically. The colour of the original reaction will darken.
3.   Rinse, counterstain, dehydrate, clear and mount as usual.

Gold chloride

Reagents
(a)   0.05% aq. gold chloride

Method
1.   Flood the preparation with the gold chloride solution.
2.   Monitor the reaction microscopically for adequate darkening.
3.   Rinse, counterstain, dehydrate, clear and mount as usual.

A.8.2   During development

Imidazole (Straus, 1982)

Reagents
(a)   Imidazole (Sigma I 0250 or I 0125)
(b)   PBS/TBS
(c)   Hydrochloric acid, 1.0 M
(d)   Standard DAB/H_2O_2 incubating solution

Method
1.   Dissolve imidazole in buffer to give a solution of 70 mg ml^{-1}.
2.   Adjust the pH to 7.0 with HCl.
3.   Add 1 ml to 100 ml of standard DAB incubating solution (final concentration of imidazole is 0.01 M).
4.   Incubate preparations as usual.

Cobalt (Hsu and Soban, 1982)

Reagents
(a) Cobalt chloride, 0.5% aq.
(b) DAB, 50 mg% in buffer
(c) Hydrogen peroxide, 30%

Method
1. Add 1 ml cobalt chloride solution to 100 ml DAB solution with stirring.
2. Incubate preparations for 5 min.
3. Mix in 10 μl H_2O_2.
4. Incubate for a further 1–5 min, checking microscopically. The reaction product is blue-black.
5. Rinse in tap water, counterstain nuclei (e.g. neutral red or methyl green), dehydrate, clear and mount in permanent medium.

Nickel (Hancock, 1982)

Reagents
(a) Nickel sulphate
(b) Acetate buffer, 0.1 M, pH 6.0
(c) DAB (50 mg aliquot in 1 ml)
(d) Hydrogen peroxide, 30%

Method
1. Soak the preparations in acetate buffer.
2. Dissolve 1–2.5 g nickel sulphate in 100 ml acetate buffer.
3. Add DAB and stir to clear the solution. Filter if necessary.
4. Add 23 μl H_2O_2 (final concentration 0.007%).
5. Incubate preparations for 5–15 min. The reaction product is dark blue-black.
6. Counterstain nuclei for 2 min in 0.1% neutral red or 2% methyl green. Rinse in tap water, blot with damp filter paper, dehydrate rapidly through two changes of 100% alcohol to solvent and mount in permanent medium.

Note: Haematoxylin is not used as a counterstain as blue nuclei do not offer sufficient contrast with the immunostain, and acid-alcohol used for differentiation can remove the nickel/DAB reaction product.

Nickel with nascent hydrogen peroxide (Shu *et al.*,1988)
Note: This method produces a very clean reaction that starts slowly and increases progressively as fresh H_2O_2 is produced in the medium from the action of glucose oxidase on glucose. Careful monitoring is required to prevent overstaining.

Reagents
(a)   Ammonium nickel sulphate hexahydrate (Fluka 09885), 2.5 g
(b)   Acetate buffer, 0.1 M, pH 6
(c)   DAB (50 mg aliquot in 1 ml)
(d) β-D-Glucose, 200 mg
(e)   Ammonium chloride, 40 mg
(f)   Glucose oxidase (Sigma G 6891), 100 U

Method
1.   Soak the preparations in acetate buffer.
2.   Dissolve the nickel salt in 100 ml acetate buffer.
3.   Add the DAB and stir to clear.
4.   Add glucose and ammonium chloride.
5.   Add the glucose oxidase
6.   Incubate the preparations for 5–20 min, checking frequently once the reaction has begun to develop. The reaction product is black and be quite intense.
7.   Finish as in the previous method for nickel.

A.9   Immunostaining methods

A.9.1   Initial procedure

Paraffin sections
1.   Cut paraffin sections at 2–5 μl (the thinner the better).
2.   Float on warm (not hot) water. If several sections are to be stained differently on the same slide, maximize the distance between the sections. Pick up on suitably coated slides (see Section A.3).
3.   Dry for several hours or overnight (longer is all right) at 37°C. If a rapid result is needed, dry the sections at 60°C for 10–15 min in an incubator (not on a hot plate) to just melt the wax. A few antigens may be damaged by this procedure. Following drying, sections may be stored indefinitely at room temperature.
4.   Remove paraffin in two changes of solvent and take through graded alcohols to water. If sections are to be isolated on the slide with a water-repellent pen, take the slide from the 100% alcohol and allow it to evaporate. Then draw round the section and, after a short drying period, replace the slide in alcohol and bring it to tap water.
5.   Block endogenous peroxidase (see Section A.4) and perform protease and/or heat-mediated antigen retrieval as required (see Sections A.5 and A.6).
6.   Place slides in buffer (PBS or TBS).

Fresh cryostat sections

1. Cut fresh sections at 5 μm from snap-frozen block of tissue and pick up on suitably coated slides (see Section A.3).
2. Allow the slides to dry at room temperature for at least 1 h and preferably longer (overnight is all right). If slides are to be stored, wrap them individually or back-to-back in 'cling-film' or aluminium foil, seal a batch of slides in a plastic bag or air-tight box containing some dessicant such as silica gel and place in a deep-freeze. Sections may be fixed before storage if preferred. Before use, allow the entire bag to warm to room temperature before opening to prevent condensation on the preparations.
3. Fix sections suitably (e.g. in 100% acetone for 10 min), then draw round sections with water-repellent pen.
4. Block endogenous peroxidase if necessary (see Section A.4).
5. Place slides in buffer (PBS or TBS).

Note: To avoid damage to the tissue structure, it is very important that cryostat sections should not become dry once they have become wet with an aqueous solution.

Pre-fixed frozen sections

1. Cut cryostat sections at 5–40 μm (depending on application), mount on coated slides (see Section A.3) and allow to dry at room temperature for at least 1 h. In some cases, frozen sections may be stored at 4°C or –20°C until used.
2. If necessary, permeabilize the sections by soaking them in a dilute solution of detergent in buffer (e.g. 0.2% Triton X-100) for 30 min. Alternatively (if the antigen permits), dehydrate the sections through graded alcohols to xylene to dissolve lipids and rehydrate via the same route (Costa *et al.*, 1980).
3. Block endogenous peroxidase and treat with protease if necessary (see Sections A.4 and A.5).
4. Place sections in buffer (PBS or TBS).

Whole-cell preparations

This procedure applies to cultured cells grown as monolayers or cytocentrifuged or allowed to settle on to (coated) slides from suspension.

1. Wash the fluid or culture medium off the preparation with buffer.
2. Fix the cells appropriately (e.g. 100% alcohol). Cell preparations may be allowed to dry after alcohol fixation or may be left in the alcohol until used. Other fixatives may require different treatment. If aqueous fixatives such as 10% formalin are used, preparations should be fixed immediately after the cells have attached to the slide. It may be possible to rinse the preparations in water and dry them for storage following fixation, but subsequently they must not be allowed to dry once they have become wet.

3. If intracellular antigens are to be immunostained and an aqueous fixative has been used, it may be necessary to permeabilize the cells with detergent. Saponin (0.1%) may be preferable to Triton X-100 for localizing some intracytoplasmic membrane antigens (Goldenthal et al., 1985).

4. Block endogenous peroxidase if necessary (see Section A.4).

5. Place preparations in buffer (PBS or TBS). At this stage, if isolation of the preparation on the slide is required, dry carefully around preparations (taking care not to let the cells become dry), draw round them with a water-repellent pen, then replace in the buffer.

Semi-thin (1 nm) epoxy resin sections (Lane and Europa, 1965)

Reagents
Saturated solution of sodium hydroxide in ethanol (impure alcohol is all right). Make the solution by adding sodium hydroxide to alcohol and shaking occasionally for several days until no more will dissolve. The solution will be light yellow-brown but will darken with time. It can be used even when dark brown and can be stored at room temperature for many weeks.

Method
1. Immerse the slides carrying the sections in the alcoholic sodium hydroxide for 10–15 min (longer may be necessary as the solution ages).

2. Rinse in three changes of 100% alcohol, for several minutes each time.

3. Check microscopically that all the resin has been removed. It is helpful to lower the condenser of the microscope for this. If necessary, replace the slide in the alcoholic sodium hydroxide for a further period.

4. If the resin has been removed, draw round the sections with a water-repellent pen, then take the slide through 90 and 70% alcohol to water, then buffer.

A.9.2   Immunostaining – all preparations

1. Take slide from buffer, shake off excess buffer and dry round preparation. If the preparation has been isolated with a water-repellent pen, ensure the hydrophobic circle is dry. If the preparation has not been isolated, wipe the slide dry except for the area of the section.

2. Place the slides on a rack in a Petri dish or other covered container with some wet paper tissue or cotton wool to maintain a humid atmosphere.

3. Background blocking: Cover each preparation completely with appropriate normal serum or protein solution diluted 1/20 in antibody diluent. Cover the incubation chamber and leave for at least 10 min (longer is all right).

4. Primary antibody. Do not rinse the slides but drain off the blocking solution on to a tissue or soak it up, leaving a minimal amount on the preparation to prevent it drying. Apply a drop of the primary antibody solution at its optimal concentration in diluent. Ensure the preparation is completely covered. Cover the dish and leave for a period of 1 h at room temperature or overnight at 4°C.
5. Rinse in three changes of buffer (PBS or TBS), for 5 min each time.
6. *(i) Direct method (labelled primary antibody)*
Fluorescent preparations: Rinse well in buffer and mount in aqueous mountant.
Enzyme preparations: Develop enzyme label. Counterstain, dehydrate and clear if appropriate and mount as required.
(ii) Indirect and three-layer methods
Dry the preparation as in step 1 and apply a drop of appropriate second antibody at its optimal dilution in diluent (no azide for enzyme conjugates). Cover the dish and leave for 30 min (or longer).
7. Rinse three times in buffer as in step 5.
8. *(i) Indirect (two-layer) method*
Fluorescent preparations: Rinse and mount in aqueous mountant.
Enzyme preparations: Develop enzyme label, counterstain, dehydrate and clear if appropriate. Mount as required.
(ii) Three-layer method
Dry round preparations as in step 1 and cover them with the appropriate third reagent at its optimal dilution (no azide for enzyme-linked reagents). Cover the dish and leave for at least 30 min.
9. Rinse three times in buffer as in step 5.
10. Fluorescent preparations: Rinse and mount in aqueous mountant.
Enzyme preparations: Develop enzyme label, counterstain, dehydrate and clear if appropriate. Mount as required.

A.10 Immunogold staining with silver enhancement (Hacker *et al.*, 1994, adapted from Holgate *et al.*, 1983)

Reagents
(a) IGSS Buffer I:
0.05 M Tris/HCl buffer, pH 7.6, containing
0.9% sodium chloride and
0.1% cold water fish gelatine (Sigma G7765)
Optional: increase concentration of sodium chloride to 2.5% and include 0.05% Tween 80 or 0.2% Triton X-100 to ensure low background. Add 0.1% sodium azide if the buffer is to be stored.

(b) IGSS buffer II:
0.05 M Tris/HCl buffer, pH 7.6, containing
0.9% sodium chloride
0.1% cold water fish gelatine
Add 0.1% sodium azide for storage
(c) Lugol's iodine: 1% iodine dissolved in 2% aq. potassium iodide
(d) 2.5% aq. sodium thiosulphate
(e) 2% glutaraldehyde in PBS, pH 7.2
For the silver enhancement
(f) Citrate buffer:
Dissolve 23.5 g of trisodium citrate dihydrate and 25.5 g of citric acid monohydrate (32.12 g anhydrous citric acid) in 850 ml distilled water
This buffer can be kept at 4°C for several weeks. Before use adjust the pH to 3.8 with citric acid solution.

Note: Solutions A and B should be prepared freshly for each run.

Solution A:
(g) Silver acetate (Fluka 85140 is recommended), 80 g in 40 ml water (double distilled if possible). Stir for 15 min to dissolve.
Solution B:
(h) Hydroquinone, 200 mg in 40 ml citrate buffer (f).

Method
1. De-paraffinize sections (draw hydrophobic circle round sections after removing the paraffin) and bring to tap water.
2. Wash in tap water for 3 min.
3. Immerse in Lugol's iodine (c) for 5 min (this oxidation step is essential, regardless of whether the fixative contained mercury).
4. Rinse briefly in tap water.
5. Immerse in sodium thiosulphate solution (d) until section becomes colourless.
6. Wash in tap water for 3 min then rinse in distilled water.
7. Immerse in IGSS buffer I (a) for 10 min, then dry slide except for the area of the section, and place in a covered humid chamber (e.g. Petri dish).
8. Apply normal serum from species providing second (gold-labelled) antibody diluted 1/10 in IGSS buffer I. Leave for 5 min.
9. Do not rinse the preparations. Drain off the serum and apply the primary antibody appropriately diluted in PBS or TBS antibody diluent. Leave for 90 min. The dilution should be determined by titration but, as a rough guide, use the dilution suitable for overnight incubation in the PAP method.
10. Wash in IGSS buffer I for 3 min then in two 3 min changes of IGSS buffer II.

11. Dry the slide, except for the area of the section and apply normal serum as in step 8 for 5 min.

12. Drain off the serum and incubate in gold-adsorbed second antibody diluted in IGSS buffer II containing 0.8% BSA for 1 h. The optimal dilution (usually between 1/25 and 1/200) should be determined by titration.

13. Rinse in 3 × 3 min changes of IGSS buffer II.

14. Post-fix in 2% glutaraldehyde (e) for 2 min.

15. Rinse briefly in two changes of distilled water (30 sec each) then in 3 × 3 min changes of distilled water. This washing must be very thorough to ensure that all halide is removed prior to silver development. Water purity is critical; since this is often variable, the use of glass-distilled deionized water may be necessary.

16. Perform silver acetate auto-metallography (see below).

17. Wash in distilled water. Fix in sodium thiosulphate (b) for 2 min.

18. Wash in running tap water for 3 min. Staining intensity may be checked microscopically. If too weak, repeat steps 15–17 with fresh silver acetate developer, using a shorter development time as there is aleady some silver on the section.

19. Counterstain as required (e.g. haematoxylin and eosin). Dehydrate, clear and mount.

A.10.1   Silver acetate auto-metallography

1. Just before use, mix solution A with solution B (see previous section).

2. Place the slides vertically in a scrupulously clean glass container and immerse in the silver amplification mixture. The colour development may be monitored microscopically and usually takes 5–20 min. If the silver solution becomes grey-black before the reaction is complete, wash the sections thoroughly and repeat the process with freshly mixed A and B.

Note 1: If non-specific silver deposits are not a problem, the silver solution may be applied as drops on the section and the process watched under the microscope.

Note 2: The final impact of an immunogold–silver stain may be further enhanced by viewing with epipolarized light. This gives extra sensitivity, allowing even single silver/gold granules to be seen by back-scattering of the light.

Note 3: Non-specific silver deposits. The inclusion of gelatine in the buffer helps to prevent these by coating the slide with a layer of gelatine. This layer, with any deposited silver, is washed off in running tap water before the preparation is counterstained.

Note 4: Slowing the silver reaction. Better control of the silver reaction may be achieved by adding gum arabic (gum acacia) to the solution. This also helps to prevent non-specific siver deposition. Dissolve the gum arabic at 500 g l^{-1} over several days with gentle stirring. Filter through several layers of gauze and store frozen in aliquots. Replace up to 60 ml of the citrate buffer (f) with an equal volume of gum arabic solution. The greater the proportion of gum arabic, the slower will be the rate of reaction.

Note 5: To remove excess silver from a preparation immerse in Farmer's solution, which consists of 90 ml 10% sodium thiosulphate and 10 ml 10 % potassium ferricyanide. *Take care:* silver is removed in seconds.

A.11   Double immunoenzymatic staining

A.11.1   Primary antibodies from different species
(adapted from Mason and Sammons, 1978)

Reagents
(a)   Two primary antibodies raised in different species (say rabbit and mouse), mixed together so that each is at its predetermined optimal dilution in antibody diluent.
(b)   Mixture of species-specific second antibodies (e.g. goat anti-rabbit Ig and goat anti-mouse Ig), each at its predetermined optimal dilution in antibody diluent.
(c)   Mixture of rabbit PAP and mouse APAAP, each at its predetermined optimal dilution in Tris buffer with 0.1% BSA but *no* azide (enzyme inhibitor).
(d)   Suitable blocking serum, in this case normal goat serum, 1/20 in antibody diluent.
(e)   Enzyme development reagents for alkaline phosphatase and peroxidase.

Note: One of the second antibodies could be biotinylated. In this case, use a peroxidase- or alkaline phosphatase-labelled streptavidin or ABC instead of the PAP or APAAP in the third layer.

Method
1.   Initial preparation of sections and any pre-treatment, including protease/heat-mediated antigen retrieval and peroxidase blocking.
2.   Block with normal serum.
3.   Apply primary antibody mixture for appropriate time.
4.   Rinse well and apply second antibody mixture for appropriate time.
5.   Rinse well, finishing with a TBS rinse, and apply third reagent mixture for an appropriate time.
6.   Rinse well in TBS.
7.   Develop alkaline phosphatase (blue or red).
8.   Rinse well in tap water. Immerse in appropriate buffer for peroxidase development.

9. Develop peroxidase in required colour. Check microscopically to stop reaction when a satisfactory colour balance with the alkaline phosphatase reaction end-point has been reached.
10. Rinse in tap water. Mount in an aqueous mountant.

A.11.2   Primary antibodies from the same species, heat-blocking method (adapted from *Lan et al.*, 1995)

Double biotin–avidin method with alkaline phosphatase and peroxidase

Reagents
(a) Two primary antibodies from different species, say mouse. Optimal dilution already established for the methods to be used.
(b) Secondary antibody (biotinylated goat anti-mouse Ig). Optimal dilution established.
(c) Alkaline phosphatase-labelled streptavidin or ABC.
(d) Peroxidase-labelled streptavidin or ABC.
(e) Blocking serum, in this case, normal goat serum, 1/20 in antibody diluent.
(f) Enzyme development reagents for alkaline phosphatase and peroxidase.
(g) Suitable buffers, diluents, mountants.

Method
1. Initial preparation of sections and any pre-treatment including protease/heat-mediated antigen retrieval, and peroxidase blocking. If only the second antibody requires heat-mediated antigen retrieval, omit this from the preliminary preparation.
2. Block with normal goat serum, 10 minutes.
3. Block endogenous biotin if necessary (Section A.4.3).
4. Apply first primary antibody for appropriate time.
5. Rinse and apply biotinylated goat anti-mouse Ig for the appropriate time, ending with a rinse in TBS.
6. Rinse and apply alkaline phosphatase-labelled streptavidin or alkaline phosphatase ABC reagent diluted in TBS diluent for the appropriate time.
7. Rinse in TBS.
8. Develop the alkaline phosphatase, e.g. blue with naphthol AS-MX phosphate and Fast Blue BB (Section A.7.2).
9. Rinse in water
10. Microwave in citrate buffer for 10 minutes (Section A.6).
11. Block with normal goat serum, 10 minutes.
12. Block spare avidin and biotin binding sites on applied labelled avidin or ABC reagent (Section A.4.3).

13.   Apply second primary antibody for appropriate time.
14.   Rinse.
15.   Apply biotinylated goat anti-mouse Ig for appropriate time.
16.   Rinse.
17.   Apply peroxidase-labelled streptavidin or ABC reagent for appropriate time
18.   Rinse.
19.   Develop peroxidase, e.g. red with AEC (Section A.7.1).
20.   Rinse.
21.   Mount in aqueous mountant.

Note 1: It is usually better to omit a blue nuclear stain which can be confused with the blue alkaline phosphatase reagent. If antigens are known to be in separate compartments, a red alkaline phosphatase and black peroxidase development could be done. In this case, a blue nuclear counterstain could be added. Use Mayer's haemalum to avoid differentiation in acid-alcohol, as the red reaction product is alcohol-soluble and the black nickel product is acid-intolerant.

Note 2: As with the previous method, PAP or APAAP reagents could be used instead of one of the labelled streptavidin or ABC methods.

Note 3: Controls: Negative controls are very important to confirm that reagents from the second reaction are not binding to reagents from the first reaction. In place of the second antibody, use an inappropriate mouse immunoglobulin or buffer diluent.

An additional control is to reverse the order of the primary antibodies. The same result should be achieved with reversed colours.

Note 4: It is usually better to do the alkaline phosphatase reaction first as the peroxidase end-product may mask antigenic sites, particularly if DAB is used.

Note 5: Two alkaline phosphatase methods could be done, one developed blue and one red, as the heat step inactivates any remaining alkaline phosphatase activity on the first reaction.

Note 6: If a double immunoperoxidase reaction is used, block any remaining peroxidase activity on the first reaction with a hydrogen peroxide step before the second reaction.

A.12   Post-embedding electron microscopical immunocytochemistry using epoxy resin-embedded tissue and an indirect immunogold method

---

Equipment
i.   Grid forceps (anti-magnetic, anti-capillary)
ii.   Wax plate (e.g. melted paraffin wax poured into a Petri dish) or

strip of Parafilm or dental wax for floating grids on drops of solution, or multiwell Terasaki plates (15 μl per well)

iii.   Millipore filters (45 μm pore size) and syringes

Reagents
(a)   10% aq. hydrogen peroxide
(b)   Saturated (approx. 5%) aq. sodium metaperiodate
(c)   0.05 M Tris/HCl buffer, pH 7.4
(d)   0.05 M Tris/HCl buffer, pH 7.4, containing 0.1% BSA (blocking solution)
(e)   0.05 M Tris/HCl buffer, pH 8.2, containing 1% BSA (diluent for gold-adsorbed antibody)

Method
Note 1: Millipore-filter all solutions.
Note 2: Wash grids with a jet from a syringe attached to a filter or (preferable because easier) immerse in drops of fluid in wells of Terasaki plates and agitate gently on a mechanical agitator, changing to fresh rinsing fluid several times over the rinsing period.
Note 3: Do not allow the grids to dry during the immunolabelling procedure.

1.   *Non-osmicated tissue.* Etch grids on drops of hydrogen peroxide (a) for 10 min to permeabilize resin.
     Osmicated tissue. Substitute sodium metaperiodate (b) for 10–30 min (check for each block) which partially reverses the fixation effects of osmium.
2.   Wash in micro-filtered distilled water and drain the grids by touching the edge with fibre-free absorbent paper.
3.   Float for 30 min on a drop of background-blocking serum (same species as second antibody, 1/20 in antibody diluent) or protein (e.g. buffer with BSA (d) for methods using Protein A–gold; Protein A binds to immunoglobulins so serum is not suitable as a blocking medium.
4.   Do not rinse, but drain the grids as in step 2 and float on or immerse in a drop of primary antibody diluted optimally in antibody diluent. Incubate for 1 h at room temperature or overnight at 4°C. The optimal dilution is usually the same as used in light microscopy.
5.   Rinse in five changes of Tris buffer (c), for 1 min each time, with agitation.
6.   Wash in three changes of Tris buffer with 0.1% BSA (d), for 1 min each time, with agitation.
7.   Transfer the grids to Tris buffer containing 1% BSA (e) for 10–15 min.
8.   Dilute the gold-adsorbed antibody in Tris buffer (e) (pH 8.0, with 1% BSA) and centrifuge at 2000 g for 10 min to remove aggregated gold particles.
9.   Incubate the grids in the supernatant for 1 h at room temperature.

10.   Rinse three times for 1 min each time in Tris buffer with 0.1% BSA, then five times for 1 min each time in Tris buffer without BSA (c), then five times for 1 min each time in distilled water.
11.   Dry the grids and contrast with lead citrate and uranyl acetate as for conventional electron microscopy. View in electron microscope.

A.13   Absorption specificity control (liquid phase)

Reagents
(a)   Antiserum at highest dilution in antibody diluent compatible with consistent labelling.
(b)   Antigen diluent: distilled water containing 1 mg ml^{-1} BSA (protecting protein) and 10 mg ml^{-1} lactose (to make small quantity of solution visible in the vial when lyophilized).
(c)   Pure (preferably synthetic) antigen in aliquots of 1 nmol per 100 μl of diluent (b), lyophilized to avoid over-dilution of the antibody. For storage, seal lyophilized vials under vacuum and place at –20°C.

Method
Consider approximate volume required to cover the sections. Warm vials of antigen to room temperature before opening. Open vials slowly, releasing the vacuum carefully.
1.   Add 100 μl diluted antibody to a vial containing 1 nmol of antigen (final concentration is 10 nmol antigen ml^{-1} of antibody). Replace the cap of the vial and shake vigorously to dissolve all the contents that may be still around the cap and sides. Re-cover (seal with parafilm) to prevent evaporation and leave at 4°C overnight or for several hours to equilibrate.
2.   Apply to test preparations in place of the primary antibody. Use unabsorbed antibody at the same dilution as a positive control and include positive control sections for both absorbed and unabsorbed antibody to show that all is in order.
3.   The level of 10 nmol ml^{-1} is usually in excess of a sub-optimally diluted antibody but if higher antigen concentrations are required, add 50 μl of antibody to 1 nmol of antigen instead of 100 μl (20 nmol ml^{-1} final concentration).
4.   To show the effect of reducing the concentration of antigen in the antibody, include a series of 10-fold diminutions of antigen concentration. Add 100 μl of buffer to a vial containing 1 nmol of antigen. Take 10 μl of this soution and add it to 100 μl of the diluted antibody (approx. 0.1 nmol antigen ml^{-1} antibody). Take a further 10 μl of the

antigen–buffer solution and add it to a further 100 μl of buffer, then add 10 μl of this to another 100 μl of antibody (approx. 0.01 nmol antigen ml^{-1} antibody). Repeat the procedure to give solutions of 0.001 and 0.0001 nmol antigen ml^{-1} antibody. Leave the diluted solutions overnight as in step 1, then use in place of the unabsorbed primary antibody. At some stage of dilution, the antigen will not be in excess of the antibody and immunostaining should reappear.

Note: The use of 1 nmol aliquots of antigen allows comparison of absorption data between different antibodies, and lyophilization of the samples is convenient for storage and provides a standard preparation. Antigen solutions can be used in liquid form (in this case, the lactose may be omitted from the solution) but the dilution factor must be remembered when antibodies are added.

References

Andrew S, Jasani B. (1987) An improved method for the inhibition of endogenous peroxidase non-deleterious to lymphocyte surface markers. Application to immunoperoxidase studies on eosinophil-rich tissue preparations. *Histochem. J.* **19**, 426–430.

Bondi A, Chieregatti G, Eusebi V, Fulcheri E, Bussolati G. (1982) The use of β-galactosidase as a tracer in immunohistochemistry. *Histochemistry* **76**, 153–158.

Burstone MS. (1961) Histochemical demonstration of phosphatases in frozen sections with naphthol AS-phosphates. *J. Histochem. Cytochem.* **9**, 146–153.

Costa M, Buffa R, Furness JB, Solcia E. (1980) Immunohistochemical localization of polypeptide in peripheral autonomic nerves using whole mount preparations. *Histochemistry* **3**, 157–165.

de Jong ASH, Van Kesse-Van Vark M, Raap AK. (1985) Sensitivity of various visualisation methods for peroxidase and alkaline phosphatase activity in immunoenzyme histochemistry. *Histochem. J.* **17**, 1119–1130.

De Mey J. (1986b) The preparation and use of gold probes. In *Immunocytochemistry, Modern Methods and Applications* (eds JM Polak, S Van Noorden) Butterworth–Heinemann, Oxford, pp. 115–145.

Goldenthal KL, Hedman K, Chen JW, August JT, Willingham M. (1985) Postfixation detergent treatment for immunofluorescence suppresses localization of some integral membrane proteins. *J. Histochem. Cytochem.* **33**, 813–820.

Graham RC, Ludholm U, Karnovsky MJ. (1965) Cytochemical demonstration of peroxidase activity with 3-amino-9-ethylcarbazole. *J. Histochem. Cytochem.* **13**, 150–152.

Hacker GW, Hauser-Kronberger C, Graf A-H, Danscher G, Gu J, Grimelius L. (1994) Immunogold–silver staining (IGSS) for detection of antigenic sites and DNA sequences. In *Modern Methods in Analytical Morphology* (eds J Gu, GW Hacker). Plenum Press, New York, pp. 19–35.

Hancock MB. (1982) A serotonin-immunoreactive fiber system in the dorsal columns of the spinal cord. *Neurosci. Lett.* **31**, 247–252.

Hanker JS, Ambrose WW, James CJ et al. (1979) Facilitated light microscopic cytochemical diagnosis of acute myelogenous leukemia. *Cancer Res.* **39**, 1635–1639.

Heyderman E. (1979) Immunoperoxidase techniques in histopathology: applications, methods and controls. *J. Clin. Pathol.* **32**, 971–978.

Holgate CS, Jackson P, Cowen PN, Bird CC. (1983) Immunogold–silver staining – new method of immunostaining with enhanced sensitivity. *J. Histochem. Cytochem.* **31**, 938–944.

Hsu SM, Soban E. (1982) Color modification of diaminobenzidine (DAB) precipitation by metallic ions and its application to double immunohistochemistry. *J. Histochem. Cytochem.* **30**, 1079–1082.

Huang WM, Gibson S, Facer P, Gu J, Polak JM. (1983) Improved section adhesion for immunocytochemistry using high molecular weight polymers of L-lysine as a slide coating. *Histochemistry* **77**, 275–279.

Lan HY, Mu W, Nikolic-Paterson DJ, Atkins RC. (1995) A novel, simple, reliable and sensitive method for multiple immunoenzyme staining: use of microwave oven heating to block antibody crossreactivity and retrieve antigens. *J. Histochem. Cytochem.* **43**, 97–102.

Lane BP, Europa DL. (1965) Differential staining of ultrathin sections of epon-embedded tissue for light microscopy. *J. Histochem. Cytochem.* **13**, 579–582.

Lunn G, Sansome E. (1990) *Destruction of Hazardous Chemicals in the Laboratory.* Wiley Interscience, New York.

Mason DY, Sammons RE. (1978) Alkaline phosphatase and peroxidase for double immunoenzymatic labelling of cellular constituents. *J. Clin. Pathol.* **31**, 454–462.

Miller RT, Kubier P. (1997) Blocking of endogenous avidin-binding activity in immunohistochemistry: the use of egg whites. *Appl. Immunohistochem.* **5**, 63–66.

Miller RT, Kubier P, Reynolds B, Henry T, Turnbow H. (1999) Blocking of endogenous avidin-binding activity in immunohistochemistry; the use of skim milk as an economical and effective substitute for biotin. *Appl. Immunhistochem. Molec. Morphol.* **7**, 63–65.

Murray GI, Foster CO, Ewen SWB. (1991) A novel tetrazolium method for peroxidase histochemistry and immunohistochemistry. *J. Histochem. Cytochem.* **39**, 541–544.

Nakane PK. (1968) Simultaneous localization of multiple tissue antigens using the peroxidase-labeled antibody method: a study on pituitary gland of the rat. *J. Histochem. Cytochem.* **16**, 557–560.

Pelliniemi LJ, Dym M, Karnovsky MJ. (1980) Peroxidase histochemistry using diaminobenzidine tetrahydrochloride stored as a frozen solution. *J. Histochem. Cytochem.* **28**, 191–192.

Shu S, Ju G, Fan L. (1988) The glucose oxidase–DAB–nickel method in peroxidase histochemistry of the nervous system. *Neurosci. Lett.* **85**, 169–171.

Straus W. (1982) Imidazole increases the sensitivity of the cytochemical reaction for peroxidase with diaminobenzidine at a neutral pH. *J. Histochem. Cytochem.* **30**, 491–493.

Suffin SC, Muck KB, Young JC, Lewin K, Porter DD. (1979) Improvement of the glucose oxidase immunoenzyme technique. *Am. J. Clin. Pathol.* **71**, 492–496.

van der Loos CM. (1999) *Immunoenzyme Multiple Staining Methods.* BIOS Scientific Publishers, Oxford.

Wood GS, Warnke R. (1981) Suppression of endogenous avidin-binding activity in tissues and its relevance to biotin–avidin detection systems. *J. Histochem. Cytochem.* **29**, 1196–1204.

Index

Page numbers in italics refer to figures or tables.

Absorption control for specificity, 91,
168–169
Adherence of preparations to slides, 24
poly-L-lysine, 24, 143–144
Polysine, 24
silane (APES), 144
Vectabond, 24
Alkaline phosphatase, 2, 49, 52
blocking endogenous enzyme, 52
with heat, 110
with levamisole, 52, 153
development
blue-brown end-product, 154
blue end-product, 153–154
red end-product, *35*, 154
in multiple staining, *38–40*, 104–105,
110, 112, 164–166
mechanism of action, 52
Alkaline phosphatase anti-alkaline
phosphatase (APAAP), 75–76
in multiple staining, *40*, 104, 110,
112
3-Amino-9-ethylcarbazole (AEC), 51,
151–152
in multiple staining, *34–40*, 105, 108
Antibody
application to preparations, 68–69
characteristics, 10–11, 67–68
affinity, 10
avidity, 10
dilution, 10–11
reaction with antigen, 68
structure, 67–68
titre, 10
choice, 131
diluent, 64, 142
for immunogold methods, 118,
161–162
dilution (optimal), 65–67, 81–85
doubling, 65, 82, 142–143
in relation to background, *36,*
65–66
in relation to buffer, 63–64
in relation to sensitivity of
technique, *37*, 67

in relation to temperature, 66–67
secondary antibody, 66
labelling, 45
labels, 45–56
biotin, 55
enzyme, 49–53
fluorescent, 46–47
gold, 53–54
hapten, 55
radioisotope, 55
monoclonal, 8–9
phage display, 9–10
polyclonal, 5–6
production, 5–6
immunization, 5–6
region-specific, 6–7, 90
storage, 64–65, 142
diluted antibodies, 64–65
enzyme-labelled antibodies, 65
undiluted antibodies, 64
testing specificity, 6, 65–67, 81–91
absorption with antigen, 168–169
dot blots, 127–128
ELISA, 6, 126–127
immunocytochemical, 6
primary antibody, optimal dilution,
36, 65–67, 81–85
radioimmunoassay, 6, 125, 126
second and third layer reagents, 6,
66, 86
Western blotting, 125, 127
Antigen retrieval, 24–28, 146–150
enzyme treatment, 24–26, 146–148
chymotrypsin, 25
neuraminidase, 148
pepsin, 147–148
protease, 147
trypsin, 146–147
heat-mediated, 26–28, *33*
in electron microscopical
immunocytochemistry, 121
in fresh tissue preparations, 27
microwave oven method, 148–149
pressure cooker method, 150
washing, 24

171

Applications of immunocytochemistry
  histopathology, 129–133
    choice of antibody, 131
    control tissue, 130–131
    tips for diagnostic laboratories,
      131–133
  research, 133
  quantification of antigen, 133–136
    confocal microscopy, 134
    electron microscopical, 134
    flow cytometry and FACS, 135
    simple methods, 135–136
Autofluorescence, 61–62
Automation, 70
Avidin–biotin methods, 55, 76–78
  avidin–biotin complex (ABC), 76–77
    disadvantages, 77
    streptavidin, 77
  blocking endogenous biotin, *37*, 61,
    77–78, 146
  labelled avidin, 76
  versus PAP, *37*

Background staining, *see* Non-specific
  staining
Biotin, 55
  avidin–biotin methods, 76–78
  blocking endogenous biotin, *37*, 61,
    77–78, 146
Buffers, 63–64
  citrate, 27, 148
  for immunogold reagents, 161–162
  PBS, 63–64, 141
  TBS, 63–64, 141–142
  Tris/EDTA, 28, 149

Cell preparations, 16–17, 18–20, *34*
  adherence to slides, 24, 143–144
  antigen retrieval, 27
  fixation, 16–17, 18–20
    cultures, 19–20
    cytospins, 19, *34*
    imprints, 18–19
    smears, 18–19
  for electron microscopy, 115, *117*
4-Chloro-1-naphthol, 51, 152
  in multiple staining, 104, 108
Controls
  absorption of antibody
    with antigen, 91, 168–169
    with *poly*-L-lysine, 86
    with serum, 87
    with tissue powder, 89
  experimental, 91
    absorption control for specificity, 91,
      168–169
  in multiple immunostaining, 104, 108
  negative, 59, 82, 90
    for monoclonal antibodies, 85–86

    for polyclonal antibodies, 85
  positive, 90–91, 130–131
  testing for non-specific binding
    by primary antibody, 85–86
    by second or third reagents, 86
    due to basic proteins, 86
  testing for unwanted specific binding
    due to cross-reactions with related
      antigens, 87–89
    due to unknown tissue antigens,
      86
Cross-reactions
  in multiple immunostaining, 106–112
  with related antigens, 87–89
    testing and removal by absorption,
      87, 88–90, 168–169
  with tissue immunoglobulins, 87
  with unknown tissue antigens, 86
Counterstains
  fluorescent, 48–49
  to mask autofluorescence, 62

Definition of immunocytochemistry, 1
Detergent
  in buffers, 64
  for permeabilization, 21
  for preventing non-specific binding,
    61
Diaminobenzidine (DAB), *33–40*, 50–51,
    150–151, 156–158
  disposal, 151
  in electron microscopical
    immunolabelling, 118, *119*, 121
  enhancement of end-product after
    reaction
    with copper sulphate, *38*, 100, 156
    with gold chloride, 100, 156
    with silver salts, 101
  enhancement of end-product during
    reaction, 100–101, 156–158
    with cobalt, 101, 157
    with imidazole, 101, 156
    with nickel , *34, 38, 39*, 101,
      157–158
  in multiple staining, *34, 38–40*, 104,
    105, 106, 108, 109
  re-use, 151
  safety precautions, 50–51, 151
Dot blots
  as antibody test, 127–128
Double dilution, 65, 82, 142–143
Double immunofluorescence, *44*, 106,
  107, 108, 110
Double immunoenzymatic staining, *34,*
  35, 38–40, 103–105, 112, 164–166

Electron microscopical immunolabelling,
  115–123, 166–168
  fixation, 115–116

labels, 118–119
  colloidal gold, 118
  ferritin, 118
  nanogold, 118
  peroxidase, 118
non-embedding, 116–117
post-embedding procedure, 166–168
  diluent for gold-labelled reagents,
    119
  immunolabelling with colloidal gold,
    121, 166–168
  immunolabelling with peroxidase,
    121, *119*
  multiple labelling, 122–123
  pre-embedding, 116
  pretreatment, 120
  sectioning resin blocks, 119
processing to resin, 117–118
quantification, 134
ELISA
  as antibody test, 6, 125, 126–127
Endogenous enzyme blocking
  alkaline phosphatase, 52, 153
  peroxidase, 49–50, 144–145
    frozen sections and whole cells, 145
    paraffin sections, 144–145
Enhancement of standard methods,
  93–101, 156–158
  build-up methods, 94–96
  intensification of DAB reaction product
    after reaction
      with copper sulphate, *38,* 100, 156
      with gold chloride, 100, 156
      with silver salts, 101
    intensification of end-product during
      reaction, 100–101, 156–158
        with cobalt, 101, 157
        with imidazole, 101, 156
        with nickel, *34, 38, 39,* 101, 157–158
Enzyme labels, 49–56
  alkaline phosphatase, 2, 49, 52,
    153–154
    blocking with levamisole, 52, 153
    development, blue-brown end-
      product, 154
    development, blue end-product,
      153–154
    development, red end-product, 154
    enzyme–antibody complex (APAAP),
      75–76
    mechanism of action, 52
    in multiple staining, *38–40,*
      104–105, 110, 112, 164–166
  β-D-galactosidase, 2, *35,* 53, 155
    development, 155
  glucose oxidase, 2, 52–53
    development, 154–155
    enzyme–antibody complex, 76
  peroxidase, 2, 49–52

blocking endogenous peroxidase,
  49–50, 144–145
development, 50–52, 150–153
enzyme–antibody complex (PAP),
  73–74
Equipment, 69

Fixation and tissue preparation, 13–23
  cell preparations, 16–17, 18–20
  combination fixatives, 15–16
    Bouin's, 15–16
    periodate-lysine-paraformaldehyde
      (PLP), 16, 116
    Zamboni's, 15, 16
  cross-linking fixatives
    formaldehyde, 14–15, 116
    glutaraldehyde, 15, 116
  electron microscopical
    immunocytochemistry, 115–116
  freeze-dried tissue, 15, 22
  frozen sections, 17, 18
  paraffin-embedded tissue, 16
  precipitant fixatives, 15
    acetone, 15
    alcohol, 15
    Carnoy's, 15
  pre-fixed, non-embedded tissue, 20–21
  problems and remedies, *82–83*
    over-fixation, 24–28, *82–83*
    under-fixation, *82–83, 84*
  storage, 22–23
Fluorescent labels, 1–2, 46–48
  advantages, 46
  Alexa dyes, 48
  AMCA, 48
  Cy-dyes, 48
  counterstains, 48–49
  fluorescein, *33, 40,* 47
  phycoerythrin, 48
  rhodamine, *44,* 47
  uses, 46–47
Freeze-dried tissue, 22
Frozen tissues, 16–18, 20–21
  fresh frozen, 16–18
    fixation, 16–18
    freezing, 17
    sectioning, 17
    storage, 23
  freeze-dried, 22
    fixation, 15, 22
  pre-fixed, 20–21
    permeabilization, 21

β-D-Galactosidase, 2, 53
  development, 155
  in multiple staining, *35*
Glucose oxidase, 52–53, 76, 154–155
  development, 154–155
  enzyme–antibody complex, 76

Gold labels, 2, 53–54
  colloidal gold, 53–54
    development with silver
      intensification, *36*, 53, 163–164
    in electron microscopy, 118–119,
      120, 122–123
  immunolabelling procedure
    electron microscopy, 166–168
    light microscopy, 161–163
  nanogold, 54, 118
  uses, 54

Haptens as labels, 55
History of immunocytochemistry, 1–3

Immunofluorescence, 2, 20, 21, 46–49, 70
  advantages, 46
  counterstains, 48–49
  disadvantages, 46
  double, *44,* 47, 48, 106, 107, 108, 109
  labels, 47–48
  mountants, 46
  quantification, 133–135
    confocal microscopy, *40,* 134
    flow cytometry and FACS, 135
  uses, *33, 40,* 46–47
Immunoglobulin, *see also* Antibody
  cross-reactivity, 87
  reaction with antigen, 68
  structure, 67–68
Immunogold labelling
  for electron microscopy, 116–123,
    166–168
    multiple labelling, 122–123
  for light microscopy, 53–54,
    with silver intensification, *36,*
      53–54, 161–164
Immunoperoxidase staining, 2, *33–40,*
  49–52
  in electron microscopical
    immunolabelling, *119,* 121
  in multiple staining, *34, 35, 38–40,*
    103–112
  increased sensitivity, 94
    with TSA, 97–100
  methods (general), 67–78, 158–161
  PAP, 73–75
  saving failed reactions, 95–96
  see also Peroxidase
Immunostaining procedure, *see also*
  Methods
  equipment, 69
  immunogold labelling for electron
    microscopy, 166–167
  immunogold staining with silver
    enhancement, 161–164
  initial procedures
    fresh cryostat sections, 158
    paraffin section, 158

pre-fixed frozen sections, 159
semithin resin sections, 160
whole cell preparations, 159
methods (general), 67–78, 158–161
Intensification of peroxidase/DAB/H_2O_2
  reaction product
  after reaction
    with copper sulphate, *38,* 100, 156
    with gold chloride, 100, 156
    with silver salts, 101
  during reaction, 100–101, 156–158
    with cobalt, 101, 157
    with imidazole, 101, 156
    with nickel, *34, 38, 39,* 101, 157–158

Labelled probes (non-immunocytochemical
  uses), 136–138
  in situ hybridization of nucleic acids,
    138
  in situ polymerase chain reaction, 138
  lectin histochemistry, 138
  receptor localization, 137
Labels (see also individual entries), 1–2,
  45–46
  biotin, 55
    avidin-biotin methods, 76–78
    blocking endogenous biotin, 61,
      77–78, 146
  enzyme, 2, 49–56
    alkaline phosphatase, 52, 153–154
    β-D-galactosidase, 2, 53, 155
    glucose oxidase, 2, 52–53, 76,
      154–155
    in multiple staining, 104–106,
      164–166
    peroxidase, 2, 49–52, 73–74,
      100–101, 118, 121, 144–145,
      150–153, 156–158
  fluorescent, 1–2, 46–49
  gold, 2, 53–54, 161–163
    in electron microscopy, 118–120,
      121, 122–123, 166–168
  haptens, 55
  radioisotopes, 55

Methods
  application of antibodies to
    preparations, 68, 69
  avidin-biotin, 55, 76–79
  direct (one-step), 70–71, 160–161
  for electron microscopy, 115–123,
    166–168
  general considerations, 63–67
    antibody diluent, 64–65
    antibody dilution, 65–67
    antibody storage, 64–65
    automation, 70
    reaction sensitivity, 67, 81–91
    reaction temperature, 66

reaction time, 65–66
indirect (two-step), 72–73, 160–161
multiple staining, 104–106, 164–166
three-step
APAAP, 75–76, 160–161
avidin-biotin, 55, 76–78, 160–161
PAP, 73–74, 160–161
Monoclonal antibodies, 8–9
advantages, 8
control for non-specific binding, 85–86
dilution, 65
disadvantages, 8–9
production, 8–9
Mountant
for fluorescence, 46
for alcohol-soluble end-products, 49
Multiple immunostaining, 103–112
primary antibodies from different
species, 104–106
double immunoenzymatic staining,
35, 36, 104–105, 106, 164–166
double immunofluorescent staining,
44, 106
double immunogold in electron
microscopy, 122–123
primary antibodies from the same
species, *34, 40,* 106–112
ARK (Dako) method, 112, 123
cross-reactions blocked
immunologically, 111–112, 123
cross-reactions blocked with
formaldehyde vapour, 110
cross-reactions blocked with heat,
34, 39–40, 110–112, 165–166
elution, dissolving reaction product,
108
elution, leaving reaction product,
107, 108, 109
masking first reaction, *34, 39,*
109–110
primary and secondary antibodies
combined *in vitro,* 112, 123
triple immunostaining, *39,* 106
quadruple immunostaining, *39*

Non-specific and unwanted staining,
59–62, 71, 72–73, 82–83, 86–90
antibody factors, 86–90
binding to basic proteins, 86–87
binding (specific) to unknown
antigens, 86
cross-reaction (specific) with tissue
Ig, 87
cross-reaction (specific) with related
antigens, 87–89
remedies, 89–90
prevention, 60–81, *82–83,* 89
absorption of antibody with tissue
powder, 89

affinity purification, 89
antibody dilution, 61–62, 89
autofluorescence, 61–62
blocking endogenous biotin, 61,
77–78, 146
blocking endogenous enzyme, 51,
49–50, 144–145, 153
blocking with normal serum or
protein, 61, 71, 72–73, 89
detergent, 52, 61
problems and solutions, *82–83*
tissue factors, 59–62
charged sites, 60
Fc receptors, 60
hydrophobic attraction, 6
Normal serum
dilution, 85
for absorbing cross-reacting Ig, 87
for blocking non-specific binding, 61,
71, 72–73, 89
for negative control, 85, 90

Paraffin-embedded tissue
adherence to slides, 24
antigen retrieval, 24–28, 146–150
fixation, 16
immunostaining, 156
sectioning, 16
storage, 22
Permeabilization of fixed tissue sections,
21
Peroxidase, 2, 49–52
blocking endogenous enzyme, 49–50,
144–145
development, 50–52, 150–153
AEC as chromogen, *34–43,* 51,
151–152
4-chloro-1-naphthol as chromogen,
51, 152
DAB as chromogen, *33–40,* 50–51,
150–151, 156–158
phenol tetrazolium method, 51, 153
in electron microscopy, 118, *119,* 121
in multiple staining, *35, 36, 34, 40,*
104–105, 106–112, 123, 164–166
intensification of DAB reaction
product, 100–101, 156–158
with cobalt, 101, 157
with copper sulphate, *38,* 100, 156
with gold chloride, 100, 156
with imidazole, 101, 156
with nickel, *34, 38, 39,* 101, 157–158
with silver salts, 101
Peroxidase anti-peroxidase (PAP)
method, 73–74
versus ABC, *37*
Phage display antibodies, 9–10
Phenol tetrazolium reaction for
peroxidase, 51, 153

Poly-L-lysine
  as absorption control, 86
  as section adhesive, 24, 143–144
Pre-fixed tissue
  frozen sections, 20–21
    initial immunostaining procedure,
      159
  permeabilization, 21, 159, 160
  Vibratome sections, 21
  whole mounts, 21
    initial immunostaining procedure,
      159–160
Problems and remedies, *82–83*

Quantification, 133–136
  confocal microscopy, 134–135
  electron microscopy, 134
  flow cytometry and FACS, 135
  simple methods, 135–136
    supra-optimal dilution, 135–136

Radioimmunoassay
  as antibody test, 6, 125, 126
Radioisotopes
  as labels, 55

Saving failed reactions, 95–96, 132–133
Sensitivity, 67, 93, 93–100
  increasing sensitivity, 93–100
    build-up methods, 94–95
    tyramine signal amplification (TSA),
      97–100, 122
  intensification of DAB reaction
    product, 100–101, 156–158
Specificity, 81–91
  non-specific and unwanted specific
    staining, 86–89
    remedies, 89–90
  testing a primary antibody, 81–85
    absorption control, 91, 168–169
  testing second and third reagents, 86
Storage
  antibodies, 64–65
  tissue preparations, 22–23
Streptavidin, 77

Tissue preparation, 13–24
  cell preparations, 18–20
    adherence to slides, 24

fixation, 16–17
permeabilization, 21
storage, 23
for electron microscopical
    immunocytochemistry, 115–118,
    119–121
  fixation, 115–116
  non-embedded, 116
  pre-embedding
    immunocytochemistry, 116
  processing to resin, 117–118
  sectioning resin blocks, 119–120
freeze-dried tissue, 22
fresh frozen tissue, 17–23
  fixation, 18
  freezing, 17–18
  sectioning, 17–18
fixation, 13–17
  combination fixatives, 15–16
  cross-linking fixatives, 14–15
  for cell preparations, 17, 18–20
  for electron microscopy, 15, 115–116
  frozen sections, 17, 18
  precipitant fixatives, 15
  vapour fixation (for freeze-dried
    tissue), 22
pre-fixed, non-embedded tissue, 20–21
  frozen sections, 20–21
  permeabilization, 21
  Vibratome sections, 21
  whole mounts, 21
paraffin-embedded tissue, 16
  adherence to slides, 24
  fixation, 16
  sectioning, 16
  storage, 22
Tyramine signal amplification (TSA),
    97–100
  for electron microscopical
    immunocytochemistry, 122

Unspecific staining, *see* Non-specific
  staining

Vibratome sections, 21

Western blotting, 125
  as antibody test, 127
Whole mount preparations, 21